Transfer and Storage of Energy by Molecules

Volume 1
Electronic Energy

Transfer and Storage of Energy by Molecules

(A Multi-volume Treatise)

Additional Volumes to Follow

Transfer and Storage of Energy by Molecules

Volume 1
Electronic Energy

Edited by

George M. Burnett *Professor of Physical Chemistry, University of Aberdeen*

Alastair M. North *Professor of Physical Chemistry, University of Strathclyde, Glasgow*

WILEY — INTERSCIENCE

A division of John Wiley & Sons Ltd

London — New York — Sydney — Toronto

First published 1969 John Wiley & Sons Ltd. All rights
reserved. No part of this book may be reproduced by
any means, nor transmitted, nor translated into a
machine language without the written permission of
the publisher.

Library of Congress catalog card number 77–78048

SBN 471 12430 3

Printed in Great Britain by J. W. Arrowsmith Ltd., Bristol 3

Preface

An understanding of the ways in which energy can be stored by, and transferred between molecules is a basic prerequisite for the successful pursuit of a wide variety of scientific disciplines. The separate developments in, for example, molecular biochemistry, solid state physics and chemical kinetics, which a decade ago seemed to be following quite divergent lines, have now reached a point where a common area of interest is clearly defined.

Of course an understanding of the various energy states available to molecules, or to systems of reacting molecules, has been the province of spectroscopists and thermodynamicists for many years. In these disciplines, however, emphasis has been placed more on the energy differences between, and the equilibrium populations of these states, rather than upon the rates and mechanisms by which the molecules move from one state to another. At the other extreme high energy or activated states have been postulated as determining the rates of chemical reactions for almost a century. Yet it is relatively recently that the question has been asked, 'How do these activated molecules obtain their energy in thermal collisions?' A rather similar state of affairs exists in the discipline of molecular biochemistry. Here, too, the importance of 'high energy' compounds in various biological processes is well established, but the mechanism of energy transfer in reaction cycles such as photosynthesis is still obscure.

In view of the considerable interest in molecular energy transfer processes there would seem to exist a definite need for a series of authoritative articles delineating the scope and setting down the fundamentals of this new discipline. It is to this end that this series of volumes has been conceived.

The publication of a number of volumes on a subject such as this immediately raises the question as to how the material should be divided. Since it is hoped that this series will form basic reference works in the field of energy storage and transfer, we have decided to subdivide the material as far as possible in terms of the form of energy involved. Thus volume 1 deals with electronic energy, volume 2 with molecular vibrational energy and volume 3 with rotational energy. It is planned that these volumes will be followed by two others dealing with energy storage and transfer processes in the solid state and in systems of biological interest.

Although the series is quite distinct from treatises on chemical kinetics or spectroscopy, it follows from the very nature of the subject that a spectroscopic or kinetic approach will be more evident in certain articles than in others. Thus in volume 1, dealing with the transfer and storage of electronic energy, the emphasis is placed on processes occurring during (or causing) chemical reactions. Chapter 1 discusses the transfer in gas phase systems and reactions, while Chapter 3 describes the chemistry of electronically excited organic molecules. The important connexion between electronic energy transfer and chemical change is thus established at the very outset.

Very often electronic energy changes involve the largest amounts of energy which can be accommodated by molecules. Consequently the relevant activation of the molecules can be achieved only at very high temperatures or under the influence of high energy radiation. Such extreme conditions are encountered in shock waves or under ionizing radiation, and are discussed in Chapters 2 and 4.

The burden of authorship is heavy when an existing young discipline is awaiting active pursuit. The editors gratefully acknowledge, therefore, the considerable amount of time and effort which the authors have devoted to make these volumes possible.

Contributing Authors

BRADLEY, J. N. *Professor of Chemistry, University of Essex, Colchester.*

BURTON, M. *Director of the Radiation Laboratory, University of Notre Dame, Notre Dame, Indiana.*

CUNDALL, R. B. *Senior Lecturer in the Department of Chemistry, University of Nottingham, Nottingham.*

FUNABASHI, K. *Member of Staff at the Radiation Laboratory, University of Notre Dame, Notre Dame, Indiana.*

HENTZ, R. R. *Member of Staff at the Radiation Laboratory, University of Notre Dame, Notre Dame, Indiana.*

KEARWELL, A. *Lecturer, Norwich City College, Norwich.*

LUDWIG, P. K. *Member of Staff at the Radiation Laboratory, University of Notre Dame, Notre Dame, Indiana.*

MAGEE, J. L. *Member of Staff at the Radiation Laboratory, University of Notre Dame, Notre Dame, Indiana.*

MOZUMDER, A. *Member of Staff at the Radiation Laboratory, University of Notre Dame, Notre Dame, Indiana.*

WILKINSON, F. *Reader in Chemistry, University of East Anglia, Norwich.*

Glossary of Symbols

The following are spectroscopic atomic state symbols

a^5D_3, a^5D_4, 4^2D, $4^2D_{5/2}$, $4^2D_{3/2}$, 6^2D_3, 6^3D_1, 6^3D_3, 7^3S_1, 7^3D_1, 8^3D_1, 2^1P, 2^3P, 3^2P, $3^2P_{1/2}$, $3^2P_{3/2}$, 4^3P_0, 4^3P_1, 4^3P_2, $5^2P_{1/2}$, $5^2P_{3/2}$, 6^3P, 6^3P_0, 6^3P_1, 1^1S_0, 2^1S_0, 2^2S, 2^3S_1, 6^1S_0, $7^2S_{1/2}$, $9^2S_{1/2}$

The following are spectroscopic molecular state symbols.

$^1A_{1g}$, ^1A_u, $A^2\Sigma^+$, $A^3\Sigma_u^{\ +}$, $A^3\Sigma$, $^1B_{2u}$, $^3B_{1u}$, $C^2\Pi$, $1D_{2u}$, $^1\Delta_g$, $^3E_{1u}$, $^n\pi^*$, 4π, $^3O_n^{\ -}$, $^1\Sigma_g^{\ +}$, $^3\Sigma_u^{\ +}$, $^3\Sigma_g^{\ -}$, $X_\pi^{\ 2}$, $X^1\Sigma_g^{\ +}$

A_i	number of atoms of atomic number Z_i in a molecule
A_{ml}	probability of emission
B	radiochemical stopping number
B_{lm}	Einstein's coefficient of absorption
c	the velocity of light
C_A, C_s	time independent coefficients in the time dependent wave function
C_v	molar heat at constant volume
C_{vib}	vibrational specific heat
D	diffusion coefficient
e	the charge on an electron
F_0	fluorescence intensity in the absence of added gas
F_x	fluorescence intensity in the presence of added gas
f_n	effective number of electrons in each atom, oscillator strength
$f_{D(\bar{v})}$	spectral distribution of fluorescence of donor
G	radiochemical efficiency
g_l, g_u	statistical weights of lower and upper states
H	Hamiltonian for a system
$h, \hbar$	Planck's constant
H'_{IJ}	matrix element for energy transfer
$\mathbf{H}_{\text{so}}$	Hamiltonian operator for spin-orbit perturbation
I	intensity of radiation (Chapter 3), mean excitation potential (Chapter 4)

I_D	intensity of delayed fluorescence
k	an orientative factor
k^0	radiative rate constant
K_D	rate constant for collisional deactivation
k_F	rate constant for fluorescence
k_{ISC}	rate constant for intersystem crossing
k_q	specific rate of quenching
K_{SV}	the Stern–Volmer constant
k_T	summation of vibrational energy transfer constants
$k_{v=n}$	rate constants for transfer from the nth vibrational level
m	the mass of an electron
$\mathbf{M}_{lm}$	the transition moment
M_M	mass of atom M
n	refractive index
N_l	number of molecules in a lower energy level
N_u	number of molecules in an upper energy level
P	concentration of free radicals in radiochemical studies (Chapter 4), density of electromagnetic radiation of frequency v (Chapter 3)
R	range of an incident particle, interaction radius
r_0	spur-size parameter
$\mathbf{r}_A$	position vector referring to electrons on acceptor
$\mathbf{r}_D$	position vector referring to electrons on donor
$\mathbf{r}_i$	position vector of the ith particle
S_{IJ}	overlap integral
T_{vib}	vibration temperature
v	velocity of a particle
$\overset{\bullet}{V}_{mn}$	matrix element of the perturbation between two states during transition
w_i	probability that the ith vibration level is initially occupied
Z	atomic number
z	the number of charges on an elementary particle
α	accommodation coefficient
ΔE	transition energy
δ	density correction
ε	extinction coefficient
$\varepsilon(v)$	molar decadic extinction coefficient at wave number v
$\varepsilon_{A(v)}$	molar decadic extinction coefficient of an acceptor
ε_n	energy of atom in state n above the ground state

η	solvent viscosity
θ_p	fraction of triplet molecules which phosphoresce
κ	dipole oscillator strength, dielectric constant
λ	de Broglie wave length
μ_e	dipole moment of the excited state
μ_g	dipole moment of the ground state
v	frequency of radiation
v_f	frequency of fluorescent emission
σ_A	quenching cross-section for mercury
σ_M	quenching cross-section of atom M
σ_n	cross section for an atom in a state n
τ	lifetime of some state or process
τ_R	radiative lifetime
ϕ	quantum efficiency of some process (Chapter 3), time dependent wave function (Chapter 4)
ϕ_{M_1}	wave function for the ground state of molecule M_1
ϕ'_{M_1}	electronic wave function for the excited state molecule M_1
ϕ_E	quantum efficiency of emission
ϕ_{isom}	efficiency of isomerization
ϕ_{triplet}	efficiency of triplet formation
ψ	wave function for a particular electronic state (Chapter 3) orientation factor (chapter 4)
ω	angular frequency

Contents

1

Electronic and Vibrational Energy Transfer in Gas Phase Systems

R. B. Cundall

1.1 INTRODUCTION

The interchange of excitation energy is an essential condition for most chemical reactions. Normal thermal reactions involve vibrational energy. In reactions induced by electric discharges, light and high-energy radiation electronic excitation provides the conditions required for many processes. The two types of energy exchange, electronic and vibrational, are usually classified as separate areas of study. Although convenient, this is unsatisfactory in some respects for gas-phase photochemistry, where activated molecules must often be regarded as in vibronic states in which the distinction between vibrational and electronic contributions to the total energy is lost.

Energy transfer considerations show the inadequacy of the usual treatment of systems by the conventional Arrhenius description of rate processes. The latter is based on a statistical treatment of the problem, whereas energy transfer research concentrates attention on the behaviour of states which are not usually present in an equilibrium Boltzmann distribution.

Electronic excitation transfer was first clearly recognized in gas-phase studies of atomic systems[1]. Analogous processes in polyatomic systems have only recently been established. Electronic excitation transfer in solution appears to be a simpler problem, since vibrational excitation is so rapidly dissipated, that it does not require consideration over the time scale of even the fastest transfer process.

Owing to the extensive area covered by the subject under discussion, this article will concentrate on some problems associated with electronic energy transfer. Vibrational effects are discussed only in so far as they

1

relate to this subject. Vibrational energy transfer and the effects arising therefrom have been extensively discussed in specialized monographs[2].

1.2 ENERGY TRANSFER IN ATOMIC SYSTEMS

Electronic energy transfer phenomena were originally recognized in systems involving excited atoms, and the principles established have been successfully extended to photochemical effects in molecular systems[1]. Mercury vapour is the most convenient for experimental study and a vast literature dealing with its photochemical behaviour has accumulated over nearly half a century[3,4]. Consequently, much of the discussion in this section deals with the $Hg(6^3P_1)$ state, which is conveniently produced by absorption of the 2537 Å resonance line by mercury vapour. Less work has been done on the excited states of metals such as cadmium, zinc and sodium, which are sufficiently volatile to act as photosensitizers[3,5]. The inert gas atoms have excited states in which there is sure to be increasing interest[6]. Research is now being done on a number of other excited atoms, e.g. arsenic[7]. Very reactive atoms such as hydrogen and oxygen merit further examination from this aspect.

Since, apart from electronic excitation, atoms can only possess translational energy, the forms of transfer are limited to (a) electronic–translation, (b) electronic–vibration, and (c) electronic–electronic. It seems most logical and instructive to follow Callear[8] and consider the results and their interpretation according to the type and amount of energy transferred. It is very apparent that the nature of the different types of transfer process is not fully understood and there is ample scope for research in this field.

1.2.1 The Quenching of the (6^3P_1) and (6^3P_0) States of Mercury

It is necessary to discuss briefly the measurement of quenching cross-sections of excited atoms. For example, $\sigma_A{}^2$ for quenching of the mercury (6^3P_1) phosphorescence by a gas, M is defined by

$$k_Q = \sigma_A{}^2\left[8\pi RT\left(\frac{M_{Hg}+M_M}{M_{Hg}M_M}\right)\right]^{1/2}$$

This is based on the mechanism

$$Hg(6^1S_0) + h\nu(2537 \text{ Å}) \rightarrow Hg(6^3P_1)$$

$$Hg(6^3P_1) \rightarrow Hg(6^1S_0) + h\nu(2537 \text{ Å})$$
$$\text{(radiative lifetime} = 1\cdot1 \times 10^{-7} \text{ s)}$$

$$Hg(6^3P_1) + M \xrightarrow{k_Q} Hg(6^1S_0) \text{ or } (6^3P_0) + A$$

The experimental techniques are discussed in detail by Mitchell and Zemansky[9]. Low pressures are needed to avoid Lorentz broadening of the emission and radiation imprisonment (or diffusion) is an effect which has made it very difficult to obtain unambiguous absolute quenching cross-sections. Zemansky[10] used thin layers of gas and viewed emissions from the back face of the cell. The results were treated using Milne's theory[11] of radiation diffusion. Changes of emission frequency introduce error and Samson[12] recalculated values given by Zemansky and Mitchell, making allowance for this. This procedure probably leads to more accurate values. Cvetanovic[4] has performed a useful service in presenting a table of quenching cross-sections in which the values and method of derivation are clearly stated. A factor of $1\cdot4$–$1\cdot45$ relates the two sets of values. Table 1.1 presents a selection of quenched cross-sections for the $Hg(6^3P_1)$ and Hg (6^3P_0) states.

Matland[13] has carried out experiments with Hg–N_2 mixtures and measured the decay time of the imprisoned radiation using the modern theory of Holstein[14]. σ^2 for nitrogen is $0\cdot42$ Å^2, a value less than that of both Samson and Zemansky. More recently Yarwood, Strausz and Gunning[15] have carried out measurements of absolute quenching cross-sections, using the Holstein theory, which yield lower values than those by Procedure II.

As well as the physical methods, relative quenching cross-sections can be measured by competitive chemical methods[16], some of which are also given in Table 1.1. The photosensitized decomposition of nitrous oxide can be used to compete with the quenching process and ϕ_{N_2} (quantum yield of N_2) can be measured for a series of different $[M]/[N_2O]$ ratios. Other chemical reactions, e.g. the *cis–trans* isomerization of butene-2[17], could also be used. Chemical methods have the advantage that they are independent of the radiation imprisonment effect, although they do not give absolute values.

Complications associated with the physical method for determining quenching cross-sections apply to atoms other than mercury.

Table 1.1

Quenching cross-sections $(\sigma_M{}^2)$ $(Å^2)$ for the deactivation of $Hg(6^3P_1)$ and $Hg(6^3P_0)$

Gas	$Hg(6^3P_1)$ physical [a] $\sigma_M{}^2(Å^2)$		$Hg(6^3P_1)$ chemical [a] $\sigma_M{}^2(Å^2)$		$Hg(6^3P_0)$ [b] $\sigma_M{}^2(Å^2)$
	Proc. I	Proc. II	Proc. I	Proc. II	
H_2	6·01	8·60			0·018
O_2	13·9	19·9			0·093
N_2	0·192	0·274			9×10^{-6}
H_2O	1·00	1·43			0·0066
D_2O	0·46	0·66			0·0048
NO	24·7	35·3	23	33	0·34
N_2O	—	—	12·6	18·0	0·51
NH_3	2·97	4·20			0·0033
CO	4·07	5·82			0·028
CO_2	2·48	3·54			0·0014
CH_4	0·059	0·085			0·007
C_2H_6	0·11	0·16	0·10	0·14	0·011
C_3H_8	1·6	2·3	1·2	1·7	0·16
C_2H_4	26	37	22	31	0·6
C_3H_6	32	46	29·8	42·6	
C_4H_8-2	39	56	39	56	
1,3 butadiene	36	51			
C_2H_2	23	33			
C_6H_6	41·9	60			

[a] Data taken from compilation of Cvetanovic[4]. Proc. I—Theory of Zemansky[10] used in treating quenching data. Proc. II—Treatment due to Samson[12].

[b] Data taken from Callear and Norrish[19] and Callear and Williams[20].

Other values based on more recent quenching experiments are given by Yarwood, Strausz and Gunning[15]. Different theoretical treatments are compared in this reference. Recent data have also been obtained by Yang[55].

1.2.2 Electronic–Vibration and Electronic–Translation Energy Transfer

Transfer of Small Amounts of Energy

One of the simplest types of electronic energy transfer is associated with reorientation of electronic spin with respect to electronic orbital angular momentum, e.g.

$$Na(3^2P_{3/2}) + Ar \rightarrow Na(3^2P_{1/2}) + Ar$$

for which the energy splitting is only 17 cm^{-1}. The spin–orbit coupling is so weak that the change is efficiently induced by collision with an inert gas atom[1,9]. The spin orbit relaxation of mercury

$$Hg(6^3P_1) + M \rightarrow Hg(6^3P_0) + M$$

has been the subject of much investigation. The energy splitting 1767 cm^{-1} is larger than for the sodium (^3P) states, and so the transfer should not occur so readily. There has been much confusion as to which molecules quench the $Hg(6^3P_1)$ state to the (6^1S_0) and (6^3P_0) states. Early attempts were made to correlate the quenching cross-section with the energy of spin–orbit splitting and the fundamental vibration frequency of the quenching molecule[9]. It has been suggested that most molecules quench directly to the (6^1S_0) state[9]. The $Hg(6^3P_0)$ state has a long radiative lifetime (10^{-3} s) since $\Delta J = 0$ is more restrictive than $\Delta S = 1$. The (6^3P_0) state can be detected by its property of ejecting electrons from a suitable metal electrode[18]. The reports of substantial amounts of the (6^3P_0) state in systems where it is now thought to be present only in low yields may be due to the fact that other excited states and free radicals affect this detector.

Much clarification emerged from the flash photolysis studies of Callear and coworkers[19,20]. The mercury 2537 Å absorption line was broadened by the addition of inert gas to give adequate light absorption from the continuous light emission of the flash lamps. The build-up and decay of the (6^3P_0) atoms in the presence of added gases was made by studying the absorptions at 4047 Å $(6^3P_0 \rightarrow 7^3S_1)$, 2967 Å $(6^3P_0 \rightarrow 6^3P_1)$, 2535 Å $(6^3P_0 \rightarrow 7^3D_1)$ and 2378 Å $(6^3P_0 \rightarrow 8^3D_1)$.

Contrary to much earlier supposition[1,4] it was found that metastable atoms were detected only in the presence of N_2, CO, H_2O and D_2O (none of which are capable of being decomposed or undergoing electronic transitions with the energy available). None were found in the presence of NO, H_2, O_2, CO_2, N_2O, CH_4, C_2H_6, C_3H_8, C_2H_4, NH_3 or BF_3. No correlation between the vibrational level separation and the energy transferred is apparent, and it appears that some form of attractive interaction must be responsible for the selective deactivation. It had been found earlier[4] that the quenching cross-sections for the quenching of the (6^3P_1) state by CO and N_2 were 4·07 and 0·19 Å^2, a difference attributed to resonance quenching mechanism. Scheer and Fine[21], by observing the ejection of electrons from a silver electrode, found that efficiencies for partial quenching of the $^3P_0$ state are about equal (CO is about 1·3 times more effective than N_2) so that a high total quenching by CO of the $Hg(6^3P_1)$ must be involved. Callear's experiments show clearly that the metastable (6^3P_0) atoms are not readily deactivated in comparison with the (6^3P_1) states (see Table 1.1).

The rôle of the (6^3P_0) state in photochemistry may not be negligible. Gunning and coworkers[22] in the course of their work on deuterated paraffins have made some interesting observations. They find that quenching cross-sections as measured by physical methods are consistently higher than those obtained by chemical methods. In such cases the (6^3P_0) state mercury atoms were found to be present in high concentrations. In the case of neopentane $(\sigma_Q^2 \approx 2 \text{ Å}^2)$ the $Hg(6^3P_0)$ is in higher yields at low pressures than it is in the presence of nitrogen. This may not be unrelated to the apparent resistance of neopentane to sensitized decomposition.

The behaviour of the (6^3P_0) state has been studied to some extent. Berberet and Clark[23] observed an ionization process proportional to the square of the intensity of the exciting 2537 Å radiation. A probable mechanism for this is

$$Hg(6^3P_1) + Hg(6^3P_0) + N_2 \rightarrow Hg''_2 + N_2$$

$$Hg''_2 \rightarrow Hg_2^+ + e$$

They also examined the previously observed emission around 4850 Å which is proportional to the first power of the light intensity. This is consistent with the scheme

$$Hg(6^3P_1) + N_2 \rightarrow Hg(6^3P_0) + N_2$$

$$Hg(6^3P_0) + Hg(6^1S_0) + N_2 \rightarrow Hg'_2 + N_2$$

$$Hg'_2 \rightarrow 2Hg(6^1S_0) + h\nu \text{ (ca. 4850 Å)}$$

The behaviour of the state has also been studied by Kimbell and Le Roy[24]. The participation of dimeric species is remarkably similar to excimer formation in the photochemistry of aromatic hydrocarbons. The Hg'_2 molecule is in the $^3O_n^-$ state according to Mrozowski[25] and radiation emission is forbidden without perturbation by collision or cross-over to a $^3\Sigma_u^+$ state.

A few other systems in which spin–orbit relaxation occurs by conversion of energy to translation have been studied. The changes

$$Se(4^3P_0) \rightarrow Se(4^3P_1)$$

$$\rightarrow Se(4^3P_2)$$

Table 1.2[a]

Deactivation of $Se(4^3P_0) \rightarrow S(4^4P_{1 \text{ or } 2})$ (25°C)

Gas	Bimolecular rate constant (cm^3 molecule^{-1} s^{-1})	Z	ΔE cm^{-1}
Ar	$2 \cdot 4 (\pm 0.3) \times 10^{-14}$	9400	-544
CO	$1 \cdot 1 (\pm 0 \cdot 3) \times 10^{-12}$	326	-391
O_2	$1 \cdot 5 (\pm 0 \cdot 3) \times 10^{-12}$	155	-544
N_2	$3 \cdot 0 (\pm 0 \cdot 3) \times 10^{-12}$	83	-203
H_2	$3 \cdot 5 (\pm 0 \cdot 7) \times 10^{-10}$	$1 \cdot 6$	-544
N_2O	$1 \cdot 2 (\pm 0 \cdot 15) \times 10^{-10}$	$2 \cdot 1$	$+ 45$
CO_2	$1 \cdot 4 (\pm 0 \cdot 1) \times 10^{-10}$	$2 \cdot 0$	$+128$

ΔE corresponds to the minimum energy to be converted to translation.

[a] Data of Callear and Tyerman[26].

can occur from selenium atoms produced by the flash-induced pre-dissociation of CSe_2 [26]. The energies of the states are: at (4^3P_0), 2534 cm^{-1}; (4^3P_1), 1990 cm^{-1}; and (4^3P_2), 0 cm^{-1}. The system has not been fully worked out and the separate transitions have not been characterized. The results available are given in Table 1.2. By making reasonable assumptions about the vibrational levels of the diatomic molecules the data can be made to conform to an approximate law of the form

$$\log Z = A\Delta E + B$$

where ΔE is the minimum energy which must be converted to transition and Z the mean collision efficiency, as suggested by Lambert and Salter[27] (Figure 1.1). The process

$$Se(4^3P_0) + O_2 \rightarrow Se(?) + O_2$$

is abnormally fast and may be due to the occurrence of a chemical reaction. Another example of spin–orbit relaxation is one found in the flash photolysis of iron carbonyl in the presence of various gases[28]

$$Fe(a^5D_3) \rightarrow Fe(a^5D_4)$$

Populations in $J = 0, 1, 2$ and 3 states are rapidly equilibrated according to the rule $\Delta J = \pm 1$. Except for Ar, rates of relaxation are much the same as for $Se(4^3P_0)$ (Table 1.3).

Very recently[29] it has been found that $As(4^2D)$ atoms can be produced by the flash photolysis of small partial pressures of $AsCl_3$ or $AsBr_3$ in an excess

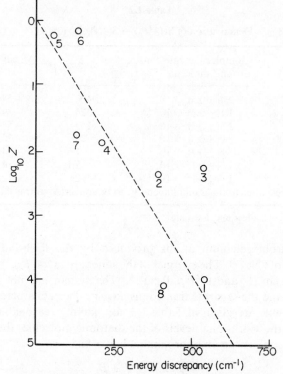

Figure 1.1 Spin–orbit relaxation of electronic ground states of $Se(4^3P_0)$ with (1) Ar, (2) CO, (3) O_2, (4) N_2, (5) N_2O, (6) CO_2, (7) is for NO (^2X) and (8) for $Fe(a^5D_3)$ $Fe(a^5D_4)$ with Ar (Callear[8])

of Ar. The process

$$As(4^2D_{5/2}) + M \rightarrow As(4^2D_{3/2}) + M, \qquad \Delta E = 323 \text{ cm}^{-1}$$

is extremely fast, as is also the total relaxation

$$As(4^2D_{3/2}) + M \rightarrow As(4^4S_{3/2}) + M, \qquad \Delta E = 10,592 \text{ cm}^{-1}$$

even for simple molecules like the inert gases (Table 1.4).

Donovan and Husain[30] have measured the rate of the spin–orbit relaxation of atomic iodine

$$I(5^2P_{1/2}) \rightarrow I(5^2P_{3/2}), \qquad \Delta E = 7603 \text{ cm}^{-1}$$

Table 1.3 [a]

Rate constants and cross-sections for the deactivation (20°c)
$$Fe(a^5D_3) \rightarrow Fe(a^5D_4)$$

Gas	Rate constant $(cm^3 s^{-1} \text{molecule}^{-1})$	Cross-section $(Å^2)$
Ar	$2 \cdot 1(\pm 0 \cdot 3) \times 10^{-15}$	$4 \cdot 1 \times 10^{-4}$
N_2	$1 \cdot 9(\pm 0 \cdot 3) \times 10^{-13}$	$3 \cdot 5 \times 10^{-2}$
He	$6 \cdot 2(\pm 0 \cdot 7) \times 10^{-14}$	$4 \cdot 7 \times 10^{-3}$
CO	$2 \cdot 9(\pm 0 \cdot 5) \times 10^{-12}$	$5 \cdot 0 \times 10^{-1}$
H_2	$7 \cdot 4(\pm 0 \cdot 7) \times 10^{-12}$	$4 \cdot 1 \times 10^{-1}$
D_2	$6 \cdot 1(\pm 0 \cdot 7) \times 10^{-12}$	$4 \cdot 7 \times 10^{-1}$
Fe	$1 \cdot 1(\pm 0 \cdot 2) \times 10^{-10}$	23

[a] Data of Callear and Oldman[28].

by the flash photolysis of I_2, CH_3I and CF_3I. The results do not follow
any simple law and may be complicated by specific chemical interaction
effects. Deactivation by I_2 is very fast, $Z = 15$, while for $N_2 \approx 10^{16}$
(Table 1.5).

Table 1.4 [a]

Rate constants and cross-sections for the deactivation of $As(4^2D_{3/2})$
$$As(4^2D_{3/2}) + M \rightarrow As(4^4S_{3/2}) + M$$

Gas	Temp. $(°K)$	Rate constant $(cm^3 \text{molecule}^{-1} s^{-1})$	cross-section $(Å^2)$
Ar	296	$1 \cdot 1(\pm 0 \cdot 2) \times 10^{-15}$	$2 \cdot 3 \times 10^{-4}$
CO	296	$4 \cdot 7(\pm 0 \cdot 6) \times 10^{-11}$	$8 \cdot 3$
CH_4	296	$1 \cdot 9(\pm 0 \cdot 4) \times 10^{-12}$	$0 \cdot 28$
CO_2	296	$7 \cdot 8(\pm 1 \cdot 2) \times 10^{-13}$	$0 \cdot 16$
Kr	296	$< 10^{-15}$	—
SF_6	296	$< 10^{-15}$	—
N_2	296	$4(\pm 0 \cdot 6) \times 10^{-12}$	$0 \cdot 72$
N_2	403	$1 \cdot 3(\pm 0 \cdot 2) \times 10^{-11}$	$2 \cdot 1$
Xe	296	$1 \cdot 7(\pm 0 \cdot 3) \times 10^{-12}$	$0 \cdot 47$
Xe	403	$5 \cdot 4(\pm 0 \cdot 8) \times 10^{-12}$	$1 \cdot 3$
H_2	296	$2 \cdot 8(\pm 0 \cdot 3) \times 10^{-11}$	$1 \cdot 6$
D_2	296	$1 \cdot 2(\pm 0 \cdot 2) \times 10^{-11}$	$0 \cdot 97$
		$As(4^2D_{5/2}) + M \rightarrow As(4^2D_{3/2}) + M$ is extremely fast.	

[a] Data of Callear and Oldman[7].

Table 1.5 [a]

Number of collisions for deactivation of
$I(5^2P_{1/2}) \rightarrow I(5^2P_{3/2})$ by various gases (27°c)

Gas	Number of collisions
He	very large
Ar	very large
CF_3I	$\sim 10^5$
N_2	$3\cdot3 \times 10^5$
C_3H_8	$3\cdot1 \times 10^3$
H_2	$5\cdot6 \times 10^3$
D_2	$3\cdot2 \times 10^3$
DI	480
HI	420
I_2	15

[a] Data of Donovan and Husain[30].

Xe is very efficient in causing deactivation in the process

$$O(2^1D) \rightarrow O(2^3P)$$

while other inert gases are inefficient[31,32]. This gives some evidence for interaction of the quenching molecule. Xe is polarizable and external spin–orbit coupling by the heavy atom may also be effective (Table 1.6).

Theoretical speculations. Two types of behaviour may be occurring: (a) a resonance process in which the P.E. curves are roughly parallel and the probability of transfer increases as the difference between the electronic energy change and the vibrational level separation decreases (increasing overlap of the translational wave functions); (b) energy transfer occurs because of crossing or close interaction of P.E. surfaces. Some correlation of efficiency with ionization potential is strongly suggestive of this type of chemical interaction (e.g. SF_6 with a high I.P. is an inefficient transfer agent).

Papers dealing with the theories of spin–orbit relaxation by collision are by Bates[33], Moskowitz and Thorson[34,35], and Bichovskii and Nikitin[36,37].

Electronic–vibration and electronic–translation energy transfer for large amounts of energy.

This type of transition corresponds to changes in orbital angular momentum or principal quantum number. Generally such highly

Table 1.6 [a]

Efficiencies of collisional deactivation of
excited states of $O(^1D_2)$ atoms (25°C)

Gas	N_2O photolysis	NO_2 photolysis
He	0·02	
Ar	—	0·011±0·06
Kr	0·09	0·060±0·005
Xe	0·82	0·78±0·02
N_2	0·26	0·24±0·02
CO_2	(1·00)	(1·00)
N_2O	2·4	1·02±0·02
NO_2	—	1·62±0·07
C_3H_8	1·74	4·67±0·33
SF_6	0	0

These values are relative to carbon dioxide.

[a] Data of Cvetanovic and Yamazaki[31] and Cvetanovic
and Preston[32].

energetic species are deactivated very efficiently by polyatomic molecules
(see Table 1.1). Callear[6] has postulated chemical complex formation to
explain the approximate correlations with polarizability and ionization
potential, which are often apparent. For the quenching of excited sodium
atoms,

$$Na(3^2P) + M \rightarrow Na(2^2S) + M$$

the inert gases have small cross-sections, whereas for polyatomic molecules the quenching cross-sections are of the same order of magnitude as
gas kinetic cross-sections[1,38]. There is no correlation between the
quenching cross-section and the minimum energy that can be taken up
as vibration (Table 1.7). In some cases selectivity is seen, as is the case for
the very efficient quenching of the $Na(3^2P)$ state by mono-olefins which
do not have a suitable low-lying energy state for electronic transfer[38].

Theoretical Models. Two relatively recent models have been proposed
for electronic–vibration transfer processes of the type being considered.
The first, due to Dickens, Linnett and Sovers[39], examines the possibility
of electronic–vibration transfer when there is no crossing of the potential
energy surfaces. Quenching cross-sections should increase with decreasing
energy discrepancy, but are very small for multiple quantum vibrational

Table 1.7[a]

Cross-sections for the deactivation of (Na^3P)

Gas	Quenching[a] cross-section $\sigma^2(\text{Å}^2)$	Vibrational level of quencher	Energy discrepancy	Quenching[b] cross-section $\sigma^2(\text{Å}^2)$flame
H_2	7·4	4	0·2	2·87
N_2	14·5	7	0·15	6·95
NO	31·6	9	0·09	—
CO	28·0	8	0·07	11·9
CH_4	0·11			
C_2H_6	0·17			
$n\text{-}C_4H_{10}$	0·3			
C_2H_4	44·0			
$C_4H_8\text{-}1$ and -2	58·0			
C_6H_6	75·0			

[a] R. G. W. Norrish and W. MacF. Smith[38]. 403°K
[b] D. R. Jenkins[59]. 1400–1900°K.

transitions due to the very small vibrational matrix elements. A specific prediction that CO should be 49 times more effective as a quencher than N_2 is made. This is only in very approximate accord with observation. It is concluded from the calculations that for quenching to be efficient there must be crossing or near crossing of the potential surfaces, i.e. strong interaction favours transfer. Also diatomic molecules should only take up a single quantum of vibrational energy. This is clearly not in accord with the experimental results. Russian workers[37] treat the transfer problem in terms of intersecting potential surfaces (i.e. a strong interaction model) and in the cases of quenching of $Hg(6^3P_1)$ and $Hg(6^3P_0)$ predict that CO and N_2 should have comparable quenching efficiencies.

These models, although crude, draw attention to the details of the processes involved, and it is now certain that chemical interaction is essential for efficient quenching of excited atoms.

Jablonski[40] first considered that electronic quenching could be considered in terms of crossing of electronic P.E. surfaces. Magee[41] and Laidler[5,42] both applied absolute reaction rate theory to the calculation of quenching rates. Laidler also drew attention to the necessity for spin-conservation[43] in the deactivation process.

Recent experimental results. Polanyi and coworkers[44,45,46] have carried out a most important series of experiments. They have attempted to measure directly the yield of vibrational energy in the quenching molecules in the $Hg(6^3P_1)$–CO and $Hg(6^3P_1)$–NO systems at low pressures by recording the infrared emission spectrum of the vibrationally excited ground state of the diatomic molecules. One, or probably both, of the reactions

$$Hg(6^3P_{1\,or\,0}) + CO(v = 0) \rightarrow CO(v = n) + Hg(6^1S_0)$$

produce vibrationally excited carbon monoxide. No resonant conversion was observed and direct excitation of CO to $v \leqslant 10$ occurred. Total quenching cross-sections differed by a factor of about ten—estimated to be between 0·09 and 1·3 Å^2.

The relative values of the rate constants for transfer (assuming $k_{v=9} = 1·00$) were

$$k_{v=2} = 80, \qquad k_{v=3} = 70, \qquad k_{v=4} = 60, \qquad k_{v=5} = 48,$$

$$k_{v=6} = 43, \qquad k_{v=7} = 35, \qquad k_{v=8} = 15, \qquad k_{v=9} = 1·00,$$

$$\text{and } k_{v \geqslant 10} \approx 0$$

The products must also have considerable amounts of translational and rotational energy ($\geqslant 53\%$ of the energy of the system).

The results for nitric oxide were very similar. There was a slow decrease in the probability of vibrational excitation of NO up to $v = 16$ or 17. Above this there is an insignificant probability of vibrational excitation. The total cross-section for electronic vibrational transfer by NO is 1–15 Å^2 for the $Hg(6^3P_1)$ state and 0·05–1·00 Å^2 for $Hg(6^3P_0)$.

For the CO quenching an explanation in terms of a HgCO* complex is proposed in which the HgCO* potential energy surface correlates with the $Hg(6^1S_0) + CO_v$. Supporting evidence for the existence of a HgCO complex comes from the experiments of Homer and Lossing[47]. They found that the addition of CO increased the rate of mercury photosensitized decomposition of ethylene, acetone and anisole by a factor of 2·5. The reaction with paraffins was unaffected by CO. Nitrogen, which produces $Hg(6^3P_0)$, was without effect. It appears that the complex HgCO, which carries most of the original excitation, can induce decomposition of some molecules. A sufficiently long lifetime would seem to be needed to make detection of such a complex possible.

It may be assumed that the similarity between the CO and NO systems is due to a participation of a HgNO complex, but there is another possible sequence in this case, namely

$$Hg(6^3P_{1\,or\,0}) + NO \rightarrow Hg(6^1S_0) + NO(^4\pi)$$

followed by

$$NO(^4\pi) \rightarrow NO(X^2\pi)$$

Gunning and coworkers[48] have marshalled considerable evidence from chemical sources to show that $Hg(6^3P_1)$ behaves towards quenchers as an electrophilic agent, i.e. it is not simply to be regarded as an energy carrier. This is illustrated by a comparison of its relative rate constant with those for $O(^3P)$[49] and $S(^3P)$[50] with the same quenchers. The order of effectiveness is $O(^3P) > S(^3P) > Hg(^3P)$.[51] Jennings and Trobridge[52] have shown that this also applies to quenching by fluoro-olefins. Gunning and Strausz[48] show that the theory is consistent with the fact that the quenching diameter of paraffinic hydrocarbons is very nearly additive (CH_3, 0·15–0·30 Å; CH_2, 1·0 Å and tertiary–CH, 1·3 Å) and the effectiveness increases with electron donating power. Superimposed on the quenching by the electrophilic influence which leads to chemical inter-action there is the possibility of resonance transfer if a suitable electronic energy state of the quencher exists, e.g. with NO and the olefins. A spin–orbit coupling effect can operate with heavy atoms.

Gunning[48] considered the structure of the transition state complex and showed that from a linear $C—H\cdots Hg$ complex a five-membered cyclic form could be produced which then gave rise to the products. This may be represented

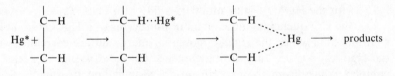

Emphasis was placed on the necessity for formation of the five-membered ring. The fact that methane and neopentane appeared to have small quenching cross-sections could be due to the fact that they cannot form five-membered ring complexes. The accepted result for neopentane is unreliable[22]. The cyclic complex mechanism has been discounted by the fact that quenching cross-sections for deuterated and light hydrogen hydrocarbons on which the theory was based are incorrect.

Gunning[53] has observed (as did workers some 40 years earlier) an emission probably due to HgM* complexes. Substrates giving rise to

$Hg(6^3P_0)$ atoms (H_2, NH_3, N_2 etc.) also produce emission from the Hg_2 excimer.

Further evidence for the formation of complexes was deduced from the apparent primary kinetic isotope effect observed when H was replaced by D. The magnitude is very close to that estimated by transition state theory. Deuterium has also been used by Gunning[48,4] to determine the site of $Hg(6^3P_1)$ attack. The appropriately labelled paraffin is decomposed in the presence of a small amount of nitric oxide or nitric oxide–olefin as radical scavengers. A mass-spectrometric technique has also been used by Lossing[54] for the same purpose. Peculiar unexplained isotope effects have been found by Lossing and Palmer[54].

Yang[55] has used the comparison of quenching rates by physical and chemical methods to determine the mechanism of interaction of 6^3P_1 mercury atoms with paraffins. Quenching by rupture of a CH bond is predominant for C_2H_6 and C_3H_8, while quenching by $C(CH_3)_4$ and $CH_3CD_2CH_3$ indicates that quenching to the metastable 6^3P_0 state is predominant. A detailed model for the electronic energy process is developed.

The technique of mono-isotopic substitution has been used to examine the nature of the primary act[48]. In this technique one of the isotopes of mercury is used in a resonance lamp and the emitted radiation specifically activates one of the natural isotopes in the reaction system. The isotopic constitution of the products is examined by mass-spectrometry. The way in which the technique is used is shown in the following example:

$$^iHg(6^3P_1) + CH_3Cl \rightarrow \,^iHgCl + CH_3$$

$$^iHg(6^3P_1) + CH_3Cl \rightarrow \,^iHg(6^1S_0) + Cl + CH_3$$

$$Cl + \,^NHg + M \rightarrow \,^NHgCl + M$$

^iHg and ^NHg are the isotopic and natural isotope mixture forms of mercury. It can be seen how the origin of calomel product can be decided from its isotopic constitution.

An alternative to the complex formation view is the hydrogen abstraction mechanism

$$RH + Hg(6^3P_1) \rightarrow HgH + R \rightarrow R + H + Hg(6^1S_0)$$

There is no evidence for HgH, but in the case of $Cd(5^3P_1)$ sensitized decompositions C—H split can only occur if there is CdH formation. This is due to the fact that the C—H bond requires more energy for rupture than can be obtained from the $Cd(5^3P_1)$ state[4].

The Reverse Vibrational–Electronic Transfer Process

Further details of the dynamics of the reactions discussed in this section may be obtained from the observation that these processes are reversible. Shock wave experiments by Gaydon and coworkers[56] show that vibrationally excited CO and N_2 will excite $Na(3^2P)$, and as a consequence of the principle of detailed balancing it must follow that vibrationally excited CO and N_2 must be formed in the reverse quenching process. Starr[57] has also confirmed the occurrence of

$$N_2(X\ ^1\Sigma_g^+ v'' > 7) + Na(3^2S) \rightarrow N_2(X\ ^1\Sigma_g^+) + Na^*$$

There has recently been a direct demonstration by the crossed beam method of direct transfer of excitation from vibrationally excited N_2 to sodium atoms by production of D-line radiation[58].

Jenkins[59] has determined the cross-sections of excited sodium atoms in flames. Cross-sections which were measured were independent of temperature. These are thought to be more reliable than resonance quenching experiments (Table 1.7) since self-absorption line broadening and compound formation are absent.

1.2.3 Electronic–Electronic Energy Transfer

Franck[60] predicted that transfer of electronic energy should occur between different electronic states of atoms and atoms of different species in their ground states. In addition he stated that as the difference in electronic excitation levels decreased, the efficiency of transfer should be enhanced. This prediction was verified by Cario and Franck[61], who irradiated mixtures of mercury and thallium with 2537 Å resonance light and observed emission from excited states of thallium just above or below the 4·86 eV of the $Hg(6^3P_1)$ level, namely the $(9^2S_{1/2})$, (6^2D_3) and $(7^2S_{1/2})$ states. Excess energy must be taken up as translational kinetic energy, or, in the case of the $(7^2S_{1/2})$ state, translational energy must become converted into electronic excitation (0·13 eV). The cross-sections for transfer were between 8 and 20 Å^2.

The formal theory has been reviewed by Massey and Burhop[62] and Mott and Massey[63]. If the change in internal energy is very small and all the transitions are optically allowed for both atoms, long range resonance transfer should result in large cross-sections for transfer of the order of 500 Å^2, but the probability should decrease rapidly, increasing energy discrepancy. The prediction that the most efficient energy transfer occurs between identical atoms has been verified by depolarization of the emitted radiation[1].

Beutler and Josephy[64] examined the mercury sensitized fluorescence of sodium and concluded that the process with the largest cross-section is

$$Hg(6^3P_1) + Na(3^2S_{1/2}) \rightarrow Hg(6^1S_0) + Na(9^2S_{1/2})$$

for which $\Delta E = 162 \text{ cm}^{-1}$ and so very little electronic energy is converted into translation. In the presence of nitrogen, which quenches $Hg(6^3P_1)$ to $Hg(6^3P_0)$, the following should also be efficient:

$$Hg(6^3P_0) + Na(3^2S_{1/2}) \rightarrow Hg(6^1S_0) + Na(7^2S_{1/2})$$

The addition of nitrogen quenched the sodium emission due to the $(9S \rightarrow 3P)$ transition but at the same time intensified the $(7S \rightarrow 3P)$ emission. This strongly supports the proposed mechanism. The intensities of other lines indicated that the $8S(\Delta E = -444 \text{ cm}^{-1})$, $7D(\Delta E = -212 \text{ cm}^{-1})$ and $8D(\Delta E = +317 \text{ cm}^{-1})$ states are populated directly by excitation transfer from mercury, the intensities of the lines being about one-fifth of the $9^2S \ 3^2P$ line. Beutler and Josephy did not record emission from the P states and there is uncertainty about the role of the process

$$Hg(6^3P_1) + Na(3^2S) \rightarrow Na(8^2P) + Hg(6^1S_0)$$

Excitation transfer from $Hg(6^3P_1)$ to indium is also consistent with the rule that energy exchange accompanied by a small internal energy change is most efficient

$$Hg(6^3P_1) + In(5^2P) \rightarrow Hg(6^1S_0) + In(7^2P)$$

$$(\Delta E = -551 \text{ and } -440 \text{ cm}^{-1})$$

$$Hg(6^3P_1) + In(5^2P) \rightarrow Hg(6^1S_0) + In(6^2D)$$

$$(\Delta E = -364 \text{ and } -314 \text{ cm}^{-1})$$

These have approximately equal cross-sections although it is significant that the first is optically forbidden and the second allowed. At 900°C, the processes

$$Hg(6^3P_1) + Tl(5^2P_{1/2}) \rightarrow Hg(6^1S_0) + Tl(6^2P)$$

$$(\Delta E = -3218 \text{ and } -3294 \text{ cm}^{-1})$$

apparently are preferred to

$$Hg(6^3P_1) + Tl(5^2P_{1/2}) \rightarrow Hg(6^1S_0) + Tl(8^2S)$$

$$(\Delta E = -666 \, cm^{-1})$$

The mercury photosensitized fluorescence of thallium provides an exception to the rule, insofar as the transfer with the largest cross-section has a substantial change in internal energy, notwithstanding the possibility of electronic excitation transfer to a state of almost identical internal energy.

Some of the complications may be understood if, as Pringsheim[1] suggested, excitation transfer may involve species other than $Hg(6^3P_1)$, e.g. Hg_2^*. Anderson and MacFarland[65] have found that complexes of the type $HgTl^*$ can produce high apparent cross-sections for transfers which would be unfavourable for resonance quenching.

Stepp and Anderson[66] suggested that there may be some conservation of electronic angular momentum accompanying transfer between atoms, and introduced the following processes:

$$Hg(6^3P_1) + Hg(6^3P_0) \rightarrow Hg(6^1S_0) + Hg(6^3D_1)$$

$$Hg(6^3P_1) + Hg(6^3P_2) \rightarrow Hg(6^1S_0) + Hg(6^3D_3)$$

Recent research on lasers provides quantitative information about excitation transfer between atoms. Javan, Bennett and Herriott[67] have observed an increase in the rate of decay of $He(2^3S_1)$ in the presence of neon and recorded a cross-section of $0.37 \pm 0.5 \, \text{Å}^2$. Apparently all four of the 2s levels of neon are populated directly, although individual cross-sections have not been reported. The internal energy changes for the processes

$$He(2^3S_1) + Ne(2^1S_0) \rightarrow He(1^1S_0) + Ne(2s_{2,3,4 \, or \, 5})$$

are -314, -469, -1053 and $-1247 \, cm^{-1}$.

The transfer from the $He(2^1S_0)$ state is predominantly[68]

$$He(2^1S_0) + Ne(2^1S_0) \rightarrow He(1^1S_0) + Ne(3S_2)$$

The recorded cross-section is $4 \, \text{Å}^2$, which is very efficient since the transition in helium is optically forbidden. Available data on excitation transfer between atoms is given in Table 1.8.

More measurements of cross-sections are required. From the limited data it is seen that $\Delta E \sim 0$ favours transfer, and optical selection rules

Table 1.8 [a]

Summary of results on excitation transfer between atoms

Donor		Acceptor		ΔE	Cross-section	Comments
Initial	Final	Initial	Final	cm^{-1}	Å^2	
$Hg(6^3P_1)$	(6^1S_0)	$Na(3^2S_{1/2})$	$(9^2S_{1/2})$	$+162$	—	F preferred to A with $\Delta E = -113\,cm^{-1}$
$Hg(6^3P_1)$	(6^1S_0)	$In(5^2P)$	(7^2P)	-440	—	F ⎱ equally
$Hg(6^3P_1)$	(6^1S_0)	$In(5^2P)$	(6^2D)	-314	—	A ⎰ probable
$Hg(6^3P_1)$	(6^1S_0)	$Tl(5^2P)$	(6^2D)	-3218	—	A
$He(2^3S_1)$	(1^1S_0)	$Ne(2^1S_0)$	$(2s_2)$	-314 ⎫		FHe ANe ⎫
$He(2^3S_1)$	(1^1S_0)	$Ne(2^1S_0)$	$(2s_3)$	-469 ⎬	0·37	FHe FNe ⎬ equally
$He(2^3S_1)$	(1^1S_0)	$Ne(2^1S_0)$	$(2s_4)$	-1053 ⎪		FHe ANe ⎪ probable
$He(2^3S_1)$	(1^1S_0)	$Ne(2^1S_0)$	$(2s_5)$	-1247 ⎭		FHe ANe ⎭
$He(2^1S_0)$	(1^1S_0)	$Ne(2^1S_0)$	$(3s_2)$	$+387$	~4	FHe ANe

F = optically forbidden transition
A = optically allowed transition

[a] Data compiled by Callear[6].

appear to have no influence on energy transfer efficiency. If there are alternative processes with ΔE of the same order the course taken cannot yet be predicted.

Transfer to Complex Molecules

For transfer to a complex molecule energy level correlation is less important, since the excess is readily converted to vibrational energy[4,5,62]. Most of the data on quenching of electronically excited atoms is for $Hg(6^3P_1)$. Possible processes are

$$Hg(6^3P_1) + RH(^1\Sigma) = Hg(6^1S_0) + RH(^3\Sigma)$$

or

$$R(^2\Sigma) + H(^2S)$$

The processes

$$Hg(6^3P_1) + RH(^1\Sigma) \left\{ \begin{array}{l} \rightarrow RH(^1\Sigma)^* \text{ (electronic excitation)} \\ \rightarrow RH(^1\Sigma)^v \text{ (vibrational excitation of the} \\ \qquad\qquad\qquad\quad \text{ground state)} \end{array} \right.$$

are not allowed due to necessity for spin conservation[5]. The sensitized isomerization and decomposition of olefins are interpreted as involving a vibrationally excited triplet state populated by triplet–triplet transfer[69]. There does not seem to be any compelling reason to think that a strong olefin–mercury complex is involved. These may be examples of molecular energy transfer at gas kinetic collision distances.

Transfer and Sensitization by Rare Gas and Other Atoms[70,71,72]

The resonance lines of the rare gases are in the vacuum ultraviolet region, where less common experimental techniques have to be used. The Xe and Kr resonance lines are at 1469·6, 1295·6, 1235·8 and 1164·9 Å respectively. Each of the states has a metastable state 0·1 eV lower, and collisional deactivation populates this level. Absolute quenching cross-sections are not known and the efficiency of transfer may vary, as is the case for $Hg(6^3P_{1\text{ or }0})$.

Electronic energy transfer from excited krypton to nitrogen leads to ammonia and hydrazine formation in the presence of hydrogen[73]. Xenon and krypton sensitize the decomposition of ethane[74]. Xenon enhances the proportion of radical decomposition over that occurring in direct photolysis. The reverse happens with krypton. The rare gas photosensitized exchange between H_2 and D_2 shows that Kr and Xe efficiently transfer to hydrogen molecules and induce dissociation into ground state atoms[75]. The yield for transfer is about unity, and this leads the authors to conclude that the quenching process is chemical. Another process established is[75]

$$Ne(^3P_{0\text{ or }1}) + O_2 \rightarrow O(2^1D) + O(2^3P) + Ne(2^1S_0)$$

Rare gas sensitized ionization (the Penning reaction)[76] occurs; anisole, NO, anthracene, ammonia and propane are ionized by excited Xe, Kr and Ar. These processes can be used for light intensity measurement in the vacuum ultraviolet.

Sensitization in the vacuum ultraviolet[70] has been accomplished by hydrogen atoms excited by Lyman α resonance light

$$H(^2S) + h\nu \rightarrow H(^2P)$$

With nitrogen these give ammonia. Another interesting type of process is[70]

$$H(^2P) + O_2 \rightarrow HO_2^+ + e$$

1.3 ELECTRONIC AND VIBRONIC ENERGY TRANSFER FROM DIATOMIC MOLECULES

Diatomic molecules in vibronic excited states are produced by absorption of light, discharge excitation and chemical reaction. Transfer of vibrational, rotational and translational energy has been very extensively studied[2]. In electronically excited states these changes must be superimposed on the effects of electronic transfer, about which there is only meagre information.

A good example of some aspects of vibronic quenching is provided by the results of Arnot and McDowell[77] with iodine at low pressures (0·02 mm). There is a resonance fluorescence originating at $v' = 25$ (not 26, as referred to before 1963) level which is produced by excitation with the 5461 Å (green) line of mercury. The transition from

$$I_2(^3\pi_{ou}^+)(v' = 25) \rightarrow I_2(^1\Sigma_g^+)(v'' = 1, 2, 3, 4, 5, 6)$$

which is shown in Figure 1.2(a) is changed by the addition of 0·5 mm of neon, due to lines which appear with $v' = 24$ or 23 as a result of loss of vibrational energy by collision (Figure 1.2b). Some anti-Stokes behaviour occurs, as shown by emission from $v' = 26$. The apparent large quenching cross-sections reported (Table 1.9) may be due to an erroneous assumption of the radiative lifetime of the $(^3\pi_{ou}^+)$ state[78]. Resonance between matching levels can evidently occur over considerable distances and vibrational changes are usually for $\Delta v = \pm 1$.

Similar experiments have been reported by Polanyi[79] and Klemperer and coworkers[80], who used a photoelectric recording technique which allowed rotational changes to be observed also.

Spin-orbit relaxation. An example of this type of transition in a diatomic molecule is provided by NO, which has $(^2\pi_{1/2})$ and $(^2\pi_{3/2})$ components[81]

Table 1.9 [a]

Coefficients for transfer of vibrational energy in excited iodine molecules ($T = 20°$c)

Gas	He	Ne	Ar	O_2
$k_T/k_F \times 10^{-4}$ litres mole^{-1}	5·31	2·69	1·64	1·5

k_T—summation of vibrational energy transfer constants
k_F—summation of fluorescence rate constants

[a] Data of Arnot and MacDowell[77].

separated by 121 cm^{-1}. The relaxation probability on collisions is 0·062 at 25°C. This is slower than for the Na($3^2P_{3/2}$) → Na($3^2P_{1/2}$) transition because of the larger amount of energy which has to be converted to translational.

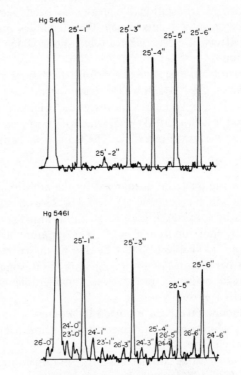

Figure 1.2 The effect of added gas on the vibrational structure of the fluorescence spectrum of iodine. (a) I$_2$, (b) effect of 0.5 mm of neon (Arnot and McDowell[77])

Electronic-electronic energy transfer. Few studies of this type of process have been made, although they must be important in discharge flow systems, particularly those involving active nitrogen[82].

One of the few cases which has been treated kinetically is the quenching of nitric oxide fluorescence by nitrogen[83]. The observed quenching of the δ-bands by nitrogen was accompanied by enhancement of the NO γ-bands, both effects showing an identical dependence on pressure. This was explained as arising from the sequential steps

$$NO(C^2\pi) + N_2(X^1\Sigma_g^+) \rightarrow NO(X^2\pi) + N_2(A^3\Sigma_u^+)$$

$$NO(X^2\pi) + N_2(A^3\Sigma_u^+) \rightarrow NO(A^2\Sigma^+) + N_2(X^1\Sigma_g^+)$$

An apparent violation of the Franck–Condon principle occurs in the first of these where there is an increase in internuclear separation of the molecules. This is possible evidence for a N_3O complex. The energy separations for these processes are -2504 and -5678 cm^{-1} in contrast to the single step

$$NO(C^2\pi) + N_2(X^1\Sigma_g^+) \rightarrow NO(A^2\Sigma^+) + N_2(X^1\Sigma_g^+)$$

for which $\Delta E = -8182$ cm^{-1}.

Small pressures of carbon monoxide caused a weakening of the γ systems due to

$$CO(X^1\Sigma^+) + N_2(A^3\Sigma_u^+) \rightarrow CO(a^3\pi) + N_2(X^1\Sigma_g^+).$$

It is not clear why the $CO(a^3\pi)$ does not excite the $NO^2(X^2\pi)$ to $NO(A^2\Sigma^+)$. An idea of the relative efficiencies of energy transfer may be obtained from Table 1.10. Sagert and Thrush[84] had earlier suggested that $N_2(A^3\Sigma_u^+)$ was an intermediate stage in converting the $(C^2\pi)$ state of NO to the $(A^2\Sigma^+)$ state.

Table 1.10 [a]

Quenching half-pressures for electronically excited NO (torr)

	$NO(X^2\Sigma)$	CO_2	$CO(X^1\Sigma^+)$
$NO(A^2\Sigma^+)$	0·91	0·31	7·1
$NO(C^2\pi)$	>15·0	20·0	28·0
$NO(D^2\Sigma^+)$		1·3	8·0

CO_2 is comparatively inefficient as a quencher of $N_2(A^3\Sigma_u^+)$ in contrast to its effect on excited NO.
CO is at least $\frac{1}{10}$ as efficient as NO in quenching $N_2(A^3\Sigma_u^+)$.

[a] Data of Callear and Smith[83].

The emission of transition metal atoms when active nitrogen reacts with metal carbonyl[85] presumably occurs by energy transfer from the

$N_2(A^3\Sigma)$ molecule to the metal atom. The transfer efficiency does not appear very dependent on the electronic state of the atom, although transfers of the singlet + triplet $\rightarrow$ singlet + singlet type are nearly 100 times less probable than spin allowed triplet + singlet $\rightarrow$ singlet + triplet type collisions.

Another case of electronic energy transfer may be associated with the blue chemiluminescence seen when iodine vapour is injected into a stream of active nitrogen[86]. Phillips[87,88] suggests that this is due to

$$I_2 + N_2(A^3\Sigma_u^+) \rightarrow N_2(X^1\Sigma_g^+) + I_2^*$$

$$I_2^* \rightarrow I_2 + h\nu \text{ (blue)}$$

The rate constant for the transfer is $(8\cdot3 \pm 1\cdot2) \times 10^{-14}$ cm^3 molecule^{-1} s^{-1}, i.e. 10^3 times less than the collision frequency. This mechanism has been strongly disputed by Young[89] since it seems that the concentration of $(A^3\Sigma_u^+)$ states is 10^{-5} times lower than required by the Phillips mechanism. The possibility of heterogeneous reaction cannot be ignored in halogen-containing systems of this type.

Transfer processes of the type described for nitrogen should also be possible in discharged oxygen. It has been shown that oxygen atoms can be generated by the reaction[90]

$$O_2(^1\Delta_g) + O_3 \rightarrow 2O_2(^3\Sigma_g^-) + O(^3P)$$

for which $\Delta E = -700$ cm^{-1}. Bader and Ogryzlo[91] find evidence for weakly bound complexes between two $O_2(^1\Delta_g)$ molecules which produce emission bands at 6340 and 7030 Å.

Some other diatomic excited molecules may act as electronic excitation energy transfer agents. Possible examples are $NO(^4\pi)$[92] and Hg^*_2[1].

1.4 ELECTRONIC AND VIBRONIC ENERGY TRANSFER FROM POLYATOMIC MOLECULES

These processes are important in photochemistry and radiation chemistry but they have only recently been characterized in the gas phase and quantitative studies are just beginning. There is no detailed information of the type we have discussed for atomic systems. The relative roles of external energy transfer, internal conversion (S–S or T–T), inter-system crossing (S–T), isomerization (or other chemical conversion) are still matters which require further research. There must be an underlying theoretical unity, but in view of the rapidly developing nature of the subject it is only possible to indicate the direction which research has taken so far. The discussion will therefore be developed by the selection of examples.

Unlike the atomic and diatomic systems considered previously, polyatomic molecules are able to change from one electronic state to another without collisional perturbation or light emission. This is regarded for the purposes of this article as intra-molecular excitation transfer. The possibilities for external energy transfer effects depend on the extent of such processes, so they will be considered in subsequent sections.

The occurrence of electronic excitation transfer by molecules in the gas phase was first shown by the observation of sensitized fluorescence from acridine and other molecules in the presence of excited naphthalene[93]. In some cases sensitized fluorescence was possible when direct excitation failed. One of the first examples of sensitized chemical excitation was reported by Dainton and Ivin[94] in 1950. Olefins quenched the photoreaction between sulphur dioxide and paraffins to produce sulphinic acids. In 1960 sulphur dioxide was found to sensitize the *cis–trans* isomerization[17] of butene-2. A triplet state of the sulphur dioxide appeared to be the activating intermediate[95,96]. The production of butene-1 from butene-2 shows that a chemical quenching process operates to at least some extent.

1.4.1 Aromatic Excitation Transfer

Dubois and Noyes[97] attempted to sensitize the dissociation of a number of molecules with excited benzene and pyridine. Only in the case of methyl iodide was success achieved, and the indications were that about every collision (within a factor of 10) between benzene and halide was effective in causing dissociation. Cundall and Palmer[17] found that electronically excited benzene caused *cis–trans* isomerization of butene-2 and postulated that triplet–triplet transfer was involved. This suggested the use of the butene-2 isomerization as a detector for triplet states in photochemical systems if the sensitizer triplet is higher in energy than the butene-2. Singlet–singlet transfer can usually be excluded since E(donor-excited singlet) $< E$(butene-2 excited singlet $\simeq 50,000\,\mathrm{cm^{-1}}$). A preliminary study of the benzene sensitized systems indicated that O_2, olefins and NO quenched excited benzene at much the same rate[98,99]. The fluorescence of benzene was unaffected by olefin and supported the suggested triplet mechanism[98]. The effect of additives discounts a free radical induced isomerization in the examples studied.

Two publications in 1959 had shown conclusively that electronically excited benzene (M) transferred energy to β-naphthylamine[100] and anthracence (F)[101]. The results are explicable by a simple competition mechanism

$$M + h\nu \rightarrow M^*$$

$$M^* + F \rightarrow F_1 + M$$

$$M^* \rightarrow M + h\nu$$

$$F_1 \rightarrow F + h\nu$$

$$M^* \rightarrow \text{radiationless decay}$$

Collisional stabilization[102] of neither excited state M^* or F_1 was required for fluorescence, and both systems probably involve singlet–singlet transfer.

The use of benzene derivatives as photosensitizers for free radical reactions is attractive in view of the complex nature of many mercury sensitized reactions. Some success in this direction was reported by Semeluk and coworkers[103]. They used benzene and hexafluorobenzene to promote decomposition of chloroform and methylene dichloride. The mechanism is unknown but the sensitizer triplet state is almost certainly not involved in the case of C_6F_6.

Noyes has made extensive contributions to benzene photochemistry[104]. Early work with Wilson[105] indicated that benzene had considerable stability to photodecomposition at wavelengths longer than 2000 Å. The fluorescence spectrum is not markedly pressure dependent above 10 torr[106]. The problems associated with benzene photochemistry are much more complex than these simple observations imply. A very detailed investigation of the fluorescence and sensitized emission from biacetyl has been made by Ishikawa and Noyes[107]. Biacetyl strongly quenches the fluorescence of benzene vapour; there is dissociation of the biacetyl (presumably from the second excited singlet state)[108]. Phosphorescence of the biacetyl is seen, but the ratio of phosphorescence (Q_P) to fluorescence (Q_F) is very large and may be infinity. The triplet state of biacetyl is produced by energy transfer from the benzene (presumably the $^3B_{1u}$ state). A Stern–Volmer plot of Q_P^{-1} vs. $[C_6H_6]$ gives a straight line, and from a limiting phosphorescence yield of 0·12 a triplet yield of 0·78 was calculated. This assumed 0·15 for the phosphorescence yield of biacetyl.

Table 1.11

The benzene–biacetyl system (irradiated at 2537 Å)

---

(1) $C_6H_6(^1A_{1g}) + hv \,(2537\,\text{Å}) \longrightarrow C_6H_6(^1B_{2u})^*$

(2) $C_6H_6(^1B_{2u})^* \xrightarrow{(+M)} C_6H_6(^1B_{2u}) \longrightarrow C_6H_6(^1A_{1g}) + hv_f$
$$k_2 = 1.6 \times 10^6 \,\text{s}^{-1}$$

(3) $C_6H_6(^1B_{2u}) \longrightarrow C_6H_6(^1A_{1g})^* \xrightarrow{(+M)} C_6H_6(^1A_{1g})$
$$k_3 = 1\text{–}1.5 \times 10^6 \,\text{s}^{-1}$$

(4) $C_6H_6(^1B_{2u}) \longrightarrow C_6H_6(^3B_{1u})^* \xrightarrow{(+M)} C_6H_6(^3B_{1u})$
$$k_4 = 3.7 \times 10^6 \,\text{s}^{-1}$$

(5) $C_6H_6(^3B_{1u}) \longrightarrow C_6H_6(^1A_{1g})^* \xrightarrow{(+M)} C_6H_6(^1A_{1g})$

(6) $C_6H_6(^1B_{2u}) + C_6H_6(^1A_{1g}) \longrightarrow 2C_6H_6(^1A_{1g})$
$$k_6 = 10^9 M^{-1}\,\text{s}^{-1}$$

(7) $C_6H_6(^1B_{2u}) + (CH_3CO)_2(^1A_{1g}) \longrightarrow C_6H_6(^1A_{1g}) + (CH_3CO)_2(^1A_u)\,(2\text{nd})$
$$k_7 = 5 \times 10^{10} M^{-1}\,\text{s}^{-1}$$

(8) $C_6H_6(^3B_{1u}) + (CH_3CO)_2(^1A_{1g}) \longrightarrow C_6H_6(^1A_{1g}) + (CH_3CO)_2(^3B_u)$

(9) $(CH_3CO)_2(^1A_u)\,(2\text{nd}) \longrightarrow \text{products}$

(10) $(CH_3CO)_2(^3B_u) \longrightarrow (CH_3CO)_2(^1A_{1g}) + hv$

(11) $(CH_3CO)_2(^3B_u) \longrightarrow (CH_3CO)_2(^1A_{1g})$

The mechanism and data obtained are summarized in Table 1.11 and the mechanism is illustrated in an energy diagram in Figure 1.3.

The conclusion drawn from this investigation was that nearly every excited singlet-state molecule which does not fluoresce undergoes intersystem cross-over. In this respect the conclusion differed from the isomerization experiments, which implied that about 15% of the excited benzene singlets internally convert to the ground state. Both sets of results have been modified by later work.

Further Work on Excited Benzene

The benzene-biacetyl and isomerization experiments illustrate difficulties associated with light emission measurements and the hazards of gas phase photochemical experiments when highly monochromatic light is not used.

Poole[109] has measured the quantum yields for fluorescence of C_6H_6 and C_6D_6 in the range 5–15 torr at 30, 60 and 90°C. A number of narrow exciting bands within the transition $^1A_{1g} \rightarrow {}^1B_{2u}$ were used. The Stern–Volmer plots indicate a slight self-quenching effect similar to that produced by cyclohexane. It is interesting to find that the fluorescence yields decrease with wavelength (2537–2345 Å) and yields for C_6D_6 exceed those

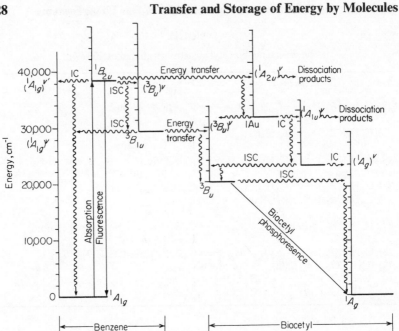

Figure 1.3 The electronic energy states of benzene and biacetyl and the transitions between them after irradiation at 2537 Å. $^1A_{1u}$ and $^1A_{2u}$ represent the first and second excited states of biacetyl (Noyes and Ishikawa[107])

for benzene under all conditions. Cross-over to the triplet state $(^3B_{1u})$ was proved by the addition of biacetyl and *cis*-butene-2, but the triplet formation also decreased with wavelength. Triplet yields for C_6H_6 and C_6D_6 produced by illumination at 2537 Å (Hg resonance lamp) are 0·70 and 0·63 (revised estimates) and lifetimes of 12·5 and 14·0 μs were deduced[110]. Noyes, Mulac and Harter[111] have submitted the measurement of emission efficiencies to a critical reassessment but have to admit that no improvement on Almy and Gillette's[112] figure of 0·15 for the biacetyl phosphorescence yield could be achieved. Three methods were used for remeasuring the benzene fluorescence yield, and a lower value of 0·18 obtained at 10 torr and 25°C (0·22 was the earlier value) is now regarded as being most reliable. No significant variation with pressure could be established, but the effect on wavelength of excitation was very marked. Below 250·0 Å it was found, in agreement with Poole, that the emission yield decreased rapidly and became essentially zero at 2400 Å and below. The new fluorescence yield when used to correct the earlier triplet yield of Ishikawa and Noyes gives 0·63. This is fortuitously close to that obtained by the isomerization energy-transfer experiments. This makes the possibility of an internal conversion process ($S_1 \rightarrow S_0$) appear more likely. Harter and Noyes[113] have verified that the fluorescence efficiency is constant from 2520–2610 Å. Both the fluorescence and triplet state yields decreased with wavelength below 2520 Å and at 2420 Å the yield may be zero. It is surprising to find that at 2535 Å a higher triplet yield is apparently found. (The corrected triplet yield value would be about one!). The results at wavelengths between 2520 and 2668 Å average to 0·70 (corrected isomerization *cis–trans* formation ratio[114a,114b]). It was shown also that the formation of benzene-butene-2 adduct was very small.

Low pressure experiments with benzene. The behaviour of excited benzene molecules at low pressures should be most informative about energy transfer and relaxation processes.

Sigal[115] has examined the isomerization of *trans*-butene-2 with C_6D_6 at total pressures below 1 torr. The relative yields of measured triplet decrease with both decreasing benzene and butene-2 pressures. If the mechanism were

$$C_6H_6 + hv \rightarrow {}^1C_6H_6$$

$$^1C_6H_6 \rightarrow C_6H_6 + hv_f$$

$$^1C_6H_6 \rightarrow {}^3C_6H_6 \tag{1}$$

$$^3C_6H_6 + trans\text{-}C_4H_8\text{-}2 \rightarrow cis\text{-}C_4H_8\text{-}2 + B \tag{2}$$

$$^3C_6H_6 \rightarrow (?) \tag{3}$$

then

$$(\phi_{isom})^{-1} = K(k^2/k^3)(P_{trans\text{-}C_4H_8\text{-}2})^{-1} + K$$

Consequently a plot of

$$(\phi_{isom})^{-1} \text{ vs. } (P_{trans\text{-}C_4H_8\text{-}2})^{-1}$$

should be linear. The data presented in Figure 1.4 shows that the behaviour is more complex. There appear to be two regions of linearity, with a transition of about 0·2 mm. Two interpretations may be offered:

(i) Intersystem crossing may have both unimolecular and bimolecular components.

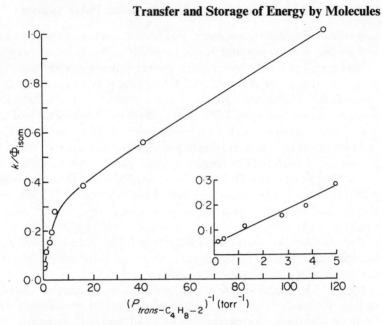

Figure 1.4 The effect of *trans*-butene-2 pressure on the quantum yield of isomerization with a constant benzene-d_6 pressure (0·060 torr). k is an arbitrary constant (Sigal[115])

(ii) Intersystem cross-over may depend on the singlet vibrational level and increase with v. This is indicated by the fact that extrapolation of the high pressure line to low pressure indicates a lower yield than is found.

Interpretation (ii) does not accord with the decrease in ϕ_T with decreasing wavelength.

Kistiakowsky and Parmenter[116] have made measurements of fluorescence and observations on triplet production at pressures where the collision frequency is low. Triplet production by energy transfer to butene-2 is slow at even the lowest pressures (0·05 torr of each gas). At 84°C the yield and intensity of fluorescent emission is constant from 0·0055 to 0·183 torr and not affected by the addition of up to 0·27 torr of benzene. At higher pressures the resonance emission spectrum changes and the yield decreases, reaching a value about 0·6 that at lower pressure when cyclohexane or CO_2 is added at pressures up to 10 torr (Figure 1.5). The low pressure yield is 0·34. It is unfortunate that the triplet yields were not measured to supplement these important observations.

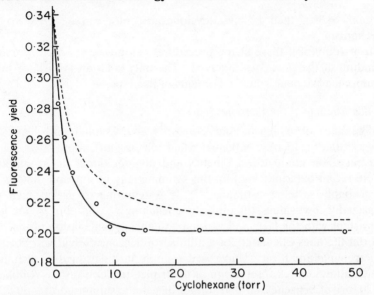

Figure 1.5   Fluorescence yield from 0·047 torr of benzene vs. pressure of cyclohexane. The experimental points are indicated, ○: the full line is the three level model; the dashed line is the two level model of Kistiakowsky and Parmenter[116]. The two level model of Strickler and Watts[117] is practically on the solid line also

The results were treated in terms of a mechanism in which there are three vibrational levels of the $^1B_{2u}$ state, each of which has the same probability of fluorescence. The best-fit curve gives relative probabilities for inter-system crossing to be 1, 1·7 and 0·5 for the lowest, intermediate and highest levels. Strickler and Watts[117] have suggested an alternative two-vibra-tional level mechanism. An S^1 level has a quantum of e_{2g} vibrational excitation which the other level S does not. The possibilities are then

$$S^1 \rightarrow k'_F$$

$$\rightarrow k'_{ISC}$$

$$\rightarrow k_D \text{ (collisional deactivation to } S)$$

$$S \rightarrow k_F$$

$$\rightarrow k_{ISC}$$

$$\rightarrow \text{ radiationless transition neglected}$$

If $k'_F = 3k_F$, then I/I_0 vs. [cyclohexane] plot can be fitted to the mechanism.

It is doubtful if these fitting procedures can give a satisfactory understanding of the processes involved. The only solution is to have more comprehensive data from experiments of this type.

The Lifetime of the Benzene Triplet

The data of Ishikawa and Noyes[107] and Cundall and Davies[110] suggest that the lifetime of the benzene triplet is short, about 10 μs. These are based on competition kinetics and are not direct determinations. More recent data indicate that this estimate may be low and that 10^{-4} s is probably a better estimate[114b]. It is tempting to assign such unexpectedly low values to impurity quenching[118], but this is not fully consistent with the facts and is not a convincing explanation of the even shorter lifetimes calculated for a number of benzene derivatives[119].

Parmenter and Ring[120] have used a simple flash sensitization technique which allows direct observation of the triplet–triplet transfer. A mixture of 20 torr of benzene and 0·01 torr of biacetyl is submitted to a 20J 3 μs flash of 2500–2600 Å light. From the time dependence of the biacetyl phosphorescence emission a lifetime of $(2·6 \pm 0·5) \times 10^{-5}$ s at 300°K can be deduced. The rate constant of $(5·7 \pm 1·1) \times 10^{11}$ cm^3 molecule^{-1} s^{-1} is obtained for the triplet–triplet transfer—a value about three times that for the corresponding singlet–singlet transfer[107].

The effect of changing from a rigid medium at $4·2 - 77°$K to the gas phase[121] must apparently increase the rate of intersystem crossing by a factor of 10^6 or some entirely new relaxation process must occur in the gas phase. Some speculations on this are made later.

Isomerization of the Benzene Ring

Photochemistry has revived interest in the isomers of benzene, Dewar benzene I, benzvalene II and prismane III.

and the possibility that these may be formed from $^1B_{2u}$ and $3B_{1u}$ states needs to be considered. This might account for some disappearance of the former and shortened lifetime of the latter.

Lipsky[122] found that benzene and alkyl benzene vapours excited in the second ($^1A_{1g} \rightarrow {}^1B_{1u}$) and third ($^1A_{1g} \rightarrow {}^1E_{1u}$) absorption bands disappear by processes other than by internal conversion to the lower $^1B_{2u}$ state since no fluorescence could be detected. Also, if the upper state undergoes intersystem crossing, then internal conversion to the $^3B_{1u}$ state cannot be efficient since biacetyl phosphorescence cannot be sensitized.

Two investigations deal with the disappearance of benzene vapour at 1849 Å. Foote, Mallon and Pitts[123] report that the quantum yield for benzene disappearance at low pressure is 0.9 ± 0.3. The major product, a valence isomer, was detected. Addition of N_2 increased the yield to a maximum at 50 mm and then reduced its production. Similar results were obtained by Shindo and Lipsky[124].

A mechanism which explains the results is

$$C_6H_6 + h\nu(1849 \text{ Å}) \rightarrow C_6H_6(^1E_{1u})$$

$$C_6H_6(^1E_{1u}) \rightarrow i - C_6H^*_6 \text{ (isomer)}$$

$$M + C_6H_6(^1E_{1u}) \rightarrow C_6H_6 \text{ (or other product than } i - C_6H_6)$$

$$i - C_6H^*_6 \rightarrow \text{(products other than } i - C_6H_6 \text{ or benzene, e.g.}$$
$$\text{polymer)}$$

The isomer formed under these conditions is fulvene[125,126] (Figure 1.6).

At 2537 Å benzvalene[127] appears to be produced, but undergoes re-aromatization to the benzene structure. This is shown by the rearrangement of benzene-1,3,5-d_3. The yield is low (~ 0.03)[128].

Excitation at shorter wavelengths leads to direct dissociation without evidence for excited states of measurable lifetime[129,130].

Benzene derivatives. The study of the effects of photoactivated benzene derivatives may, apart from their intrinsic interest, help further the understanding of the details of energy transfer effects in benzene itself. Both the biacetyl and isomerization techniques for the detection of triplet states are applicable but chemical effects may interfere, e.g. bromine atoms from the photolysis of bromobenzene can induce a chain reaction isomerization of butene-2. Fluorinated and alkyl benzenes are convenient in this respect and are most easily studied.

Methylation decreases the yield of triplet state and appears to increase the triplet lifetime[131]. In fluorobenzenes intersystem crossing probabilities may be affected by spin–orbit coupling[121].

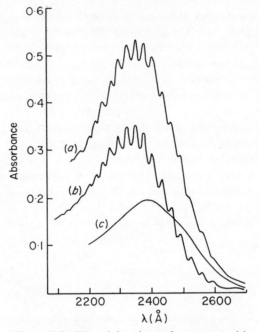

Figure 1.6 Ultraviolet absorption spectra: (a) benzene vapour irradiated at 1849 Å vs. benzene blank; (b) fulvene vapour; (c) difference spectrum $a-b$ (Kaplan and Wilzbach[126])

Unger[132] found that ϕ_f for monofluorobenzene is 0·235 at zero pressure, and the triplet yield was measured to be 0·90 at 2537 Å. Phillips[133] has shown that experimental artifacts make both yields high and that $\phi_f + \phi_{triplet} \approx 1$. The yields are the same at 2650, 2540 and 2480 Å; at 2390 Å the triplet yield is less, but can be increased by the addition of cis-butene-2.

The photochemistry of hexafluorobenzene is of particular interest because this molecule has the same symmetry properties as benzene. The triplet yield, as measured by energy transfer, is very low[134] (0·03 or less) and may even be zero[131]. The fluorescence yield is also very small ($<0·02$)[134]. Clearly the excited singlet state undergoes some very fast unimolecular process which is more effective with decreasing wavelength. Haller[135] has shown that irradiation of C_6F_6 vapour gives the Dewar form. Gases capable of removing vibrational energy enhance the iso-merization. Quantum yields of the isomerization process obtained by extrapolating the observed yields to infinite pressure are 0·0207, 0·0463 and 0·081 at 2652, 2482 and 2288 Å respectively. The triplet state is excluded because $Hg(6^3P_1)$ sensitization does not give the Dewar form and *cis*-butene-2 does not inhibit isomerization.

A number of other aromatic substances have been studied as energy transferring agents for the electronic excitation of olefins. It is found that the triplet yield is a function of the molecular symmetry[131]; it changes markedly amongst *o*-, *m*- and *p*-disubstituted benzenes for example. The yield decreases with increasing substitution on the aromatic ring. Fluorination very markedly reduces the lifetime of the triplet state[119].

The triplet yields from the $(n-\pi^*)$ and $(\pi-\pi^*)$ singlet states of pyridine are 0·14 and 0·01 or less[131,136]. Since there is no fluorescence from pyridine here again seems to be an example of a rapid conversion process.

Studies on the 2537 Å induced isomerization of *o*-xylene[131,137] shows that the extent of *m*-xylene formation is low ($\phi \approx 0.01$) and that the triplet is not a precursor. Experiments with 1,3,5-benzene-d_3[128] and mesity-lene[138] support this conclusion.

The Effect of Vibrational Energy Carry-over

Sato and coworkers[139] have carried out studies on the *cis-trans* iso-merization of dideuteroethylene with various sensitizers, benzene, toluene, *o*-, *m*- and *p*-xylene, mesitylene, monofluorobenzene and benzotrifluoride. In addition to *cis–trans* isomerization, hydrogen atom scrambling occurs in the olefin and 1,1-dideuteroethylene is formed at the lower pressures. The results require that two excited states of the olefin $C_2H_2D^*_2$ and $C_2H_2D^{**}_2$ are involved and a plausible mechanism is as follows:

$$S^* + trans - C_2H_2D_2 \rightarrow C_2H_2D^*_2 + S$$

$$C_2H_2D^*_2 \rightarrow C_2H_2D^{**}_2$$

$$C_2H_2D^*_2 + M \text{ (deactivator)} \rightarrow 0.5 \text{ } trans\text{-}C_2H_2D_2 + 0.5 \text{ } cis\text{-}C_2H_2D_2 + M$$

$$C_2H_2D^{**}_2 + M \rightarrow 0.33 \text{ } trans\text{-}C_2H_2D_2 + 0.33 \text{ } cis\text{-}C_2H_2D_2$$

$$+ 0.33 \text{ } asym\text{-}C_2H_2D_2 + M$$

Kassel's[140] equation for the reaction of a vibrationally activated molecule is used to calculate the differences in triplet state energies of the sensitizers. The procedure may be questionable since it pays insufficient regard to the varying possibilities for energy distribution on activation and differing efficiencies of subsequent deactivation.

It is an essential of the olefin activation technique for triplet measurement that transfer between olefin molecules as

$$^3C_4H_8 - 2^* + cis\text{-}C_4H_8 - 2 \rightarrow trans\text{-}C_4H_8\text{-}2 + ^3C_4H_8\text{-}2^*$$

does not occur. No evidence for this process has been found. This is almost certainly due to the fact that the triplet and ground state potential energy surfaces of the olefins intersect and intersystem crossing is very rapid (10^{-9} s). The two states considered by Sato[139] are probably different vibrational levels of the olefin ground state.

If E_T (sensitizer) $\leqslant E_T(C_4H_8 - 2)$, the transfer may require an activation energy. Naphthalene vapour excited at 3130 Å causes isomerization of butene-2 but an activation energy of about 12 kcal/mole can be calculated from the observed temperature dependency.

The Mechanism of the Triplet-Triplet Transfer Process

Haninger and Lee[114a] have made a competitive study of the efficiency of various olefin molecules in quenching the benzene triplet. The variation of *trans*-butene-2 yield from *cis*-butene-2 in the presence of other quenchers was measured (Table 1.12). The effect of alkyl substitution demonstrates that the benzene triplet is an electrophilic donor like $Hg(6\,^3P_1)$ or $O(^3P)$[142]. Oxygen is only 0·3 times as efficient as *cis*-butene-2.

This implies that some form of chemical complexing is essential for excitation transfer. This would overcome the restricting Franck–Condon condition of rotation around the C—C bond on excitation, which might be expected to apply if simple exchange interaction occurred at normal gas kinetic encounter distances.

Other benzene photosensitized reactions.

The isomerization of methyl isocyanide to methyl cyanide can be sensitized by benzene as well as by direct excitation[143]. Quantum yields of about 2 at zero pressure were explained by an unique energy chain process

$$CH_3NC^* + CH_3NC \rightarrow 2CH_3NC \ or \ CH_3CN^* + CH_3CN$$

$$CH_3CN^* + CH_3NC \rightarrow CH_3CN + CH_3NC^*$$

The complete quenching by O_2 suggests, but does not prove, that the triplet state may be involved. Reported sensitizations by N_2O and CO_2 at 2537 Å are very surprising and require further investigation.

Table 1.12 [a]

Relative rates of interactions of various olefins with triplet benzene
($^3B_{1u}$) and the triplet O(^3P) atom (23°C)

Olefin	$C_6H_6(^3B_{1u})$	O(^3P)
Ethylene	0·16±0·02	0·042
Propylene	0·51±0·04	0·24
Butene-1	0·50±0·04	0·24
Pentene-1	0·51±0·04	—
Hexene-1	—	0·27
cis-Butene-2	(1.00)	(1.00)
trans-Butene-2	1·08±0·08	1·19
Cyclopentene	1·06±0·08	1·25
Trimethylethylene	1·7±0·2	3·3
Tetramethylethylene	3·0±0·3	4·3
1,3-Butadiene	11·8±2	
O_2	0·3	

[a] Haninger and Lee[114a] and Cvetanović[142a].

Benzene photosensitization provides a convenient way of populating triplet states of the much-investigated carbonyl series. This is useful for studying the reactions of triplet states without the presence of excited singlet states. It has the advantage over mercury in that lower energies are involved and complexes of the type HgM* are unlikely to be important.

Rebbert and Ausloos[144] have compared the direct and benzene sensitized photolysis of 3-methylpentenal to give cis- and trans-butene-2 and butene-1 by intra-molecular rearrangement. The aldehyde undergoes both triplet–triplet and singlet–singlet transfer with excited benzene. 0·1 torr of aldehyde is sufficient to quench all the benzene triplet (estimated $\phi_T \simeq 0·50$). Lee[145] has carried out a similar study with 4-pentenal which is four times more efficient in quenching the benzene triplet than cis-C_4H_8-2. Both singlet and triplet transfer can occur to cyclopentonone[146] and it is calculated that the benzene triplet lifetime is rather longer than 0·3 μs. This is low and would be more appropriate to a singlet excitation transfer mechanism. Ho and Noyes[147] have shown that decomposition of ketene is sensitized by singlet and triplet excited benzene. The yield of 0·71 for benzene triplet excited at 2520 Å is in excellent agreement with that obtained by other methods. Benzene has also been used to sensitize the isomerization of crotonaldehyde[148].

1.4.2   Sensitization by Transfer from Excited States of Other Compounds

Carbonyl compounds act as sensitizers by both singlet–singlet and triplet–transfer processes[149]. An example in which these possibilities were used to elucidate a mechanism is the acetone–olefin system investigated by Cundall and Davies[150]. The transfer of excitation from the acetone triplet to the olefin was a function of the vibrational energy of the sensitizing triplet. At 48°C and 3130 Å excitation, the rate constants for

$$^3CH_3COCH_3 + C_4H_8 - 2 \rightarrow CH_3COCH_3 + {}^3C_4H_8 - 2$$

varied between 2×10^7 and $2 \times 10^9 \ M^{-1} s^{-1}$. This type of behaviour can give much useful information on the energy possessed by the triplet state. Acetaldehyde[151] did not show such a spread in transfer efficiency, only $1-2 \times 10^8 \ M^{-1} s^{-1}$. This combined with other facts leads to a suggested comparison in the photochemistry of these two compounds. The results of such a comparison are shown in Table 1.13 and Figure 1.7.

Table 1.13 [a]

Comparison of data for acetaldehyde and acetone at 48°C with 3130 Å exciting light

Datum	Acetaldehyde	Acetone
(a) Triplet state yield	0·4	1·0
(b) Dissociation from triplet	0·2	0·15
apparent activation energy	ca. 5 kcal/mole	6·4 kcal/mole
rate constant at 48°C	$6·2 \times 10^5 s^{-1}$	$1·25 \times 10^6 s^{-1}$
(c) Dissociation from singlet	Negligible at 48°C	Negligible at 48°C
(d) Non-radiative decay yield	0·1 − 0·2	0·85
from triplet rate	$4 \times 10^5 s^{-1}$	$1·25 \times 10^5 s^{-1}$
constant at 48°C		
(e) Internal conversion yield	ca. 0·6	0·0
from singlet		
(f) Fluorescence yield	5×10^{-3}	2×10^{-3}
rate constant at 48°C	—	$3·4 \times 10^5 s^{-1}$
(g) Phosphorescence yield	2×10^{-3}	2×10^{-2}
rate constant at 48°C	$8 \times 10^2 s^{-1}$	$5 \times 10^3 s^{-1}$
(h) Transfer of triplet energy to		
cis-butene-2		
Rate constants at 48°C	$1-2 \times 10^8$ l mole^{-1} s^{-1}	2×10^7 to 2×10^9 l mole^{-1} s^{-1}

[a] Data compiled by Cundall and Davies[149].

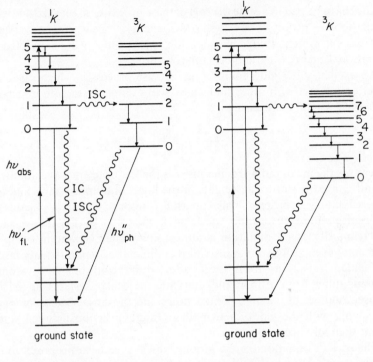

Acetaldehyde   Acetone

Figure 1.7 Effect of the vibrational energy level structure on intersystem crossing hv_{abs} absorption of one quantum, hv_{fl} fluorescence, hv_{ph}. Note ISC shown from $^1K_1$ where matching with triplet levels. Cross-over must take place from a low vibrational level (Cundall and Davies[149])

The diradical character of the $(n-\pi^*)$ states may mean that chemical interactions may be more pronounced than in aromatic excitation transfer systems.

1.4.3 Mechanisms for Electronic Relaxation of Polyatomic Molecules

In the parlance of Robinson[152,153,154] the term *electronic relaxation* covers the following phenomena: radiationless conversion between two excited states, internal conversion to the ground state, intersystem crossing between excited states and intersystem crossing to the ground state. In earlier sections it has been indicated how electronic energy transfer studies have helped to decide the role of these processes in selected systems. Radiationless processes are first order energy transfer (or conversion) processes in which electronic excitation energy after being converted to vibrational energy eventually becomes heat.

An excited state undergoing none of these processes, which does not react or interact with other molecules, will eventually radiate in accord with the Einstein relationship

$$\frac{1}{\tau} = \frac{v^2}{3\cdot47 \times 10^8} \frac{g_l}{g_u} \int \varepsilon \, \mathrm{d}v$$

where τ is the mean radiative lifetime, v is the frequency in wave numbers, g_l and g_u are the statistical weights in the lower and upper states and ε is the extinction coefficient. This equation is approximately applicable to molecules whose absorption spectrum is mirrored by the fluorescence spectrum. Resonant radiation is almost completely lacking in complex molecules even at low pressures and it is difficult to separate truly intra-molecular transitions from those affected by collisions. Polyatomic systems differ from atomic and diatomic molecular systems in the increased number of possible excited states and the close spacing of levels associated with the vibrational modes. Interconversion between states might therefore occur more easily.

Theories of radiationless transitions which have been proposed differ in the emphasis which is placed on environmental interactions, i.e. external energy transfer processes. Robinson and Frosch[153,154] have developed a theory in which the effect of the environment is specifically considered. In a solution, or crystal, the solute molecule couples to the environment which provides a large number of final states into which the initial state may degenerate. This theory accepts that the transfer is independent of the nature and concentration of the surrounding molecules. The process can be represented as follows:

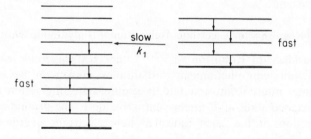

The rate-determining step is the unimolecular crossing (k_1) to a nearly isoenergetic vibrational level of the lower electronic state. Gouterman's[155] theory considers the situation in which the crossing is strongly perturbed and vibrational relaxation is slow. This is a model which is probably applicable to diatomic molecules in the gas phase at low pressures and where fluorescent transitions do not necessarily occur from the lowest vibrational level of the first excited singlet state.

Other workers have treated the relaxation process, in particular Ross[156]. He ignores environmental effects, except to assume that vibrational relaxation is very fast. Factors affecting electronic relaxation are Franck–Condon ones, i.e. problems of molecular geometry.

The Robinson theory is more applicable to polyatomic molecules, and it has been extended to gaseous molecules through a specific consideration of benzene[157]. Robinson considers two limiting situations: (a) the small-molecule limit, where the electronic relaxation cannot occur; and (b) the large-molecule limit, where electronic relaxation would be virtually the same as in solution. In other words, a molecule with a large number of vibrational degrees of freedom may provide the quasi-continuum of states needed to take up the electronic excitation converted to vibrational energy. A molecule like benzene corresponds to an 'intermediate' situation where relaxation cannot occur in the free molecule but only slight interactions are needed to induce such a process. The details of the levels in the intermediate case are inaccessible to spectroscopic resolution, and only minute perturbations are needed to produce the quasi-continuum. The $^1B_{2u}(v = 0) \rightarrow {}^3B_{1u}$ transition may correspond to the intermediate case, the separation of states in this condition being around $0.003\ cm^{-1}$. The vibrationally excited $^1B_{2u}$ state will be at the large-molecule limit, since the intramolecular motion associated with the vibration provides the necessary perturbation for relaxation. This explains why the fluorescence yield is less than unity even at zero pressure, as implied by the results of Kistiakowsky and Paramenter[116].

The theory does not allow the possibility of a unimolecular $^1B_{2u} \rightarrow {}^1A_{1g}$ transition owing to the large amount of energy which needs to undergo electronic–vibrational conversion.

Robinson discusses the possible role of the $^3E_{1u}$ state, which is thought to lie about $900\ cm^{-1}$ below the lowest $^1B_{2u}$ vibrational level[158]. The vibrational levels of the lowest vibrational level of the $^1B_{2u}$ state are assumed to be coarsely spaced around the isoenergetic cross-over level, and so mix less readily with the $^1E_{2u}$ levels than the more closely spaced $^3B_{1u}$ levels.

This neglects interaction between $^3E_{1u}$ and $^3B_{1u}$ states and collisional perturbations may make conditions less restrictive. Support for Robinson's views that the molecule can act as its own environment for taking up the energy discrepancy between the electronic states may be taken from the agreement between the fluorescent yields in the gas at pressures greater than 10 torr $(0{\cdot}18)^{111}$ and that determined in EPA glass at 77°K $(0{\cdot}2)^{159}$.

Some form of internal conversion of excited states of benzene derivatives certainly occurs in the gas phase systems, for example, for C_6F_6 $\phi_{IC} \approx 1$. A possible explanation for this type of process may be approached through the theories of Bryce-Smith and Longuet-Higgins[160] put forward to explain the chemical reactions of photoactivated benzene.

For the $^1B_{2u}$ state they proposed the following:

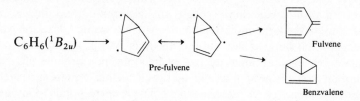

$$C_6H_6(^1B_{2u})$$

Pre-fulvene

Fulvene

Benzvalene

Fulvene and benzvalene are regarded as tautomers of the $^1B_{2u}$ state since they belong to the same symmetry class. The measured yields of these molecules in the gas phase systems are too low to account for all the molecules which do not undergo fluorescence or intersystem crossing. It is possible that the pre-fulvene intermediate may revert to the $^1A_{1g}$ state rather than fulvene or benzvalene.

In the case of the $^3B_{1u}$ state the following possibilities arise

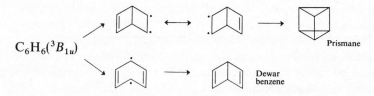

$$C_6H_6(^3B_{1u})$$

Prismane

Dewar benzene

Again the formation of the diradical intermediates may assist intersystem crossing for return to the $^1A_{1g}$ state.

In both cases the diradical intermediates have similar symmetry properties to the excited states, and the suggested scheme is very successful in correlating the chemical reactivity, although proof of the existence of these intermediates is still to be obtained. They may only need to be regarded as transition states. The production of the diradicals would be enhanced by vibrational excitation and the increase in 'internal conversion' with decreasing wavelength is consistent with this. The fact that the yield of fulvene approached unity at 1849 Å and zero pressure is also in agreement with the hypothesis.

Siebrand[161] has also developed a theory for understanding radiationless transitions. One application of the theory is the triplet ground state transition. The rate constant for the non-radiative process is

$$k = \tau_0^{-1} - k^0$$

where τ_0 is the observed triplet lifetime and k^0 is the radiative rate constant. In rigid media k^0 is about $30 \, \text{s}^{-1}$. A plot of $\log k$ against E (the triplet energy) is approximately linear: better correlations can be obtained by allowing for structural differences (Figure 1.8). Deuteration increases the triplet lifetime. Two possible explanations are: (a) a decreased probability of quantum mechanical tunnelling[162]; or (b) Franck–Condon factors are proportional to the vibrational frequencies of the modes (which are reduced by deuteration).

Some examples to illustrate the effect of deuteration are given in Table 1.14. These are not gas phase results.

1.4.4 The Lifetime of the Triplet States of Aromatic Hydrocarbons in the Gas Phase

The problems presented by the experimental observations on the benzene and similar molecule triplet lifetimes have already been discussed. A number of polyacene triplets have been investigated. They also have short lifetimes, of the order of 10^{-3} s.

Dikun[163] reported that the emission from excited phenanthrene had two components, one having the lifetime of the prompt fluorescence and the other having a longer lifetime of about 10^{-3} s. This was interpreted as meaning that the triplet phenanthrene could undergo 10^5 collisions before undergoing thermal activation to the excited singlet, which then underwent fluorescence. Williams[164] has observed long-lived emissions from anthracene, perylene, pyrene and phenanthrene vapours which he suggested were due to long-lived excited dimers which could undergo thermal dissociation to regenerate the singlet excited monomer which

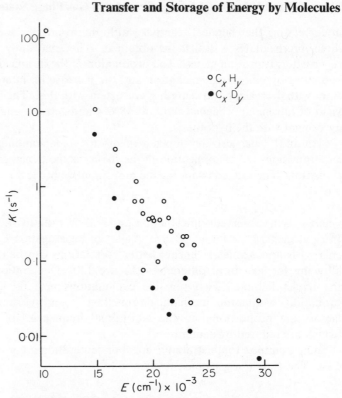

Figure 1.8 Plot of the non-radiative triplet decay constant k against the energy of lowest triplet state for normal and deuterated aromatic hydrocarbons (Siebrand[161])

produced the delayed emission. Stevens and his co-workers[165] interpreted the pressure dependence of the constant for delayed emission k_D in terms of a Lindemann mechanism for unimolecular dissociation of the dimer or excimer. The key to understanding of the phenomenon was provided by Parker and Hatchard[166] who observed that the intensity of delayed fluorescence I_D depended on the square of the absorbed light intensity. The mechanism for delayed emission is now regarded as

(a) $$A^3 + A^3 \rightarrow {}^1A + A$$

(b) $${}^3A \rightarrow A$$

(c) $${}^1A \rightarrow A + h\nu$$

Table 1.14 [a]

Energies (E) and lifetimes of the lowest triplet state of some aromatic hydrocarbons

Compound		$E(cm^{-1})$	τ_0 (s)	$k(s^{-1})$
Benzene	C_6H_6	29,500	16	0·029
	C_6D_6		26	0·0050
Naphthalene	$C_{10}H_8$	21,300	2·4	0·39
	$C_{10}D_8$		19	0·020
Anthracene	$C_{14}H_{10}$	14,700	0·045	22
	$C_{14}D_{10}$		0·14	7·1
Phenanthrene	$C_{14}H_{10}$	21,600	3·5	0·25
	$C_{14}D_{10}$		16	0·029
Pyrene	$C_{16}H_{10}$	16,900	0·5	2·0
	$C_{16}D_{10}$		16	0·29
Triphenylene	$C_{18}H_{12}$	23,300	16	0·029
	$C_{18}D_{12}$		22	0·012
Chrysene	$C_{18}H_{12}$	20,000	2·6	0·36
	$C_{18}D_{12}$		13	0·043
1,2-Benzanthracene	$C_{18}H_{12}$	16,500	0·3	3·3
	$C_{18}D_{12}$		1·7	0·59
p-Terphenyl	$C_{18}H_{14}$	20,600	2·6	0·36
	$C_{18}D_{14}$		5·3	0·16

τ_0—the observed radiative lifetime
k—non-radiative rate constant

[a] Selected values from table compiled by W. Siebrand[161a].

Under conditions of low light intensity such that

$$k_1 \gg k_2[^3A]$$

the decay constant for delayed fluorescence is given by

$$k_D = -\frac{\mathrm{d}}{\mathrm{d}t}(\ln I_D) = -\frac{2\mathrm{d}}{\mathrm{d}t}(\ln {}^3A) = 2k_1$$

Porter and Wright[167] found that the rates of the intersystem crossing $^1A \rightarrow {}^3A$ and $^3A \rightarrow A_0$ were independent of pressure. A more recent investigation[168] showed that the vapour phase decay of naphthalene and anthracene triplets are entirely explicable in terms of second order triplet–triplet quenching. This is a very efficient process with a rate close to the encounter frequency. Quenching by O_2 is rather surprisingly only about 1% of the collision rate; this is also the case for quenching of carbonyl triplets[149] (Table 1.15).

Table 1.15 [a]

Quenching rate constants for the naphthalene ($^3C_{10}H_8$) anthracene ($^3C_{14}H_{10}$) triplets

Quenching reaction	Temperature (°C)	Quenching constant ($M^{-1} s^{-1}$)	Collision rate ($M^{-1} s^{-1}$)
$^3C_{10}H_8 + O_2$	90	$2 \cdot 1 \pm 0 \cdot 4 \times 10^9$	$3 \cdot 0 \times 10^{11}$
$^3C_{10}H_8 + {}^3C_{10}H_8$	90	$6 \cdot 3 \pm 0 \cdot 8 \times 10^{10}$	$3 \cdot 6 \times 10^{11}$
$^3C_{14}H_{10} + O_2$	140	$2 \cdot 1 \pm 0 \cdot 8 \times 10^9$	$3 \cdot 5 \times 10^{11}$
$^3C_{14}H_{10} + {}^3C_{14}H_{10}$	140	$1 \cdot 4 \pm 0 \cdot 4 \times 10^{11}$	$3 \cdot 8 \times 10^{11}$

[a] Data of Porter and West[168].

Stevens[169] has used the delayed fluorescence of naphthalene, phenanthrene, anthracene, pyrene, perylene, 1,12-benzperylene, anthanthrene and fluoranthene to follow the decay of the triplet states in the vapour phase. The lifetimes were about 10^{-3} s in all cases, but they sometimes depended on the diameter of the vessel and the vapour temperature[170]. These effects were shown to be due to quenching impurities present in the solid sample, which may or may not be completely vaporized at the sample temperature. In the case of phenanthrene and fluoranthrene the delayed emissions were due to impurities which in the former case was anthracene. There was observed to be a decrease in the triplet lifetime with increase in vapour temperature at constant pressure, which is inconsistent with an increase in collisional quenching and is tentatively attributed to a dependence of the rate of radiationless relaxation of the triplet state on its vibrational energy content. The results are shown in Table 1.16.

Delayed fluorescence[171,172] has been used to measure the relative proportion γ of (ii) in the overall triplet–triplet annihilation process (i):

$$2^3A \rightarrow \text{products} (^1A + A \text{ or } 2A) \qquad \text{(i)}$$

and

$$2^3A \rightarrow {}^1A + A \qquad \text{(ii)}$$

Using the rate constant for (i) measured by Porter and West[168], γ is found to vary from 0·3 to 1·0.

1.4.5 Theories of Bimolecular Energy Transfer

Excitation transfer by the trivial reabsorption of emission is unimportant in gases at low pressures (a few torr). The long-range excitation transfer

Table 1.16 [a]

Limiting values for the triplet state decay constants in
the vapour state

Compound	$T\,°C$	$k\,(s^{-1})$
Naphthalene	93	$\leqslant 350$
Phenanthrene	230	$\leqslant 230$
Fluoranthrene	250	$\leqslant 270$
Pyrene	200	$\leqslant 180$
Perylene	230	$\leqslant 570$
1,12-Benzperylene	360	$\leqslant 940$
Anthracene	220	$\leqslant 3000$
Anthracene	340	380 ± 20

[a] Data of Stevens, Walker and Hutton[169].

mechanism discussed by Perrin[173], Förster[174] and Dexter[175] in which dipole–dipole interaction leads to coupling of the optical transitions in donor and acceptor molecules is very rare, if established at all, in gas phase transfer. This may be due to the fact that suitable molecules, such as those which undergo this type of transfer in solid and liquid media[176], have not been examined in the vapour phase. Multipole–multipole and electron exchange interactions fall off very rapidly with distance and only are effective at the molecular encounter distances which apply in gas phase systems. Some evidence for chemical interactions between donor and acceptor in specific examples has been mentioned.

1.4.6 Internal Conversion between States

Excited atoms can radiate from higher excited states. Emissions from a number of excited states are also observed from diatomic and simple polyatomic molecules at normal pressures in the gas phase. Electronic transitions can occur in such molecules, as the frequent occurrence of predissociation shows. For complex polyatomic molecules in a condensed phase or vapour, fluorescence nearly always originates from the lowest excited singlet state regardless of the excitation wavelength. The extent to which interaction with the surroundings is necessary for internal conversion between excited states is uncertain and data is very limited.

Pringsheim[177] and Stevens[178] showed that fluorescence of anthracene was essentially the same for excitation of first (3650 Å) or second (2400 Å) excited singlet states. The same effect was seen at pressures so low that most molecules suffered no gas phase collisions during the lifetime of the excited state. Similar results were obtained for naphthacene[179] and naphthalene[180]. The emission from the states produced by excitation at shorter wavelengths had a diffuse structure of the kind expected from a molecule with a large excess of vibrational energy, and Stevens noticed an appreciable red-shift.

The theories of Gouterman and Robinson are directly concerned with this type of transition, but rigorous calculations are impossible due to lack of knowledge of vibrational frequencies and anharmonicity factors. The Gouterman theory requiring a strong interaction with the environment is certainly not applicable to the three hydrocarbons investigated. They are examples of the Robinson case (b)—large molecule class.

Experiments of this type are difficult to carry out and further work needs to be done with pure samples under conditions where surface effects are eliminated or minimized. The possibility of delayed fluorescence affecting the observations cannot be ignored.

1.4.7   Excited State Stabilization

Earlier in this chapter mention was made of vibrational energy transfer from a specific vibrational energy level of the first excited electronic state of iodine. A number of other examples are available, e.g.

$$NO(A^2\Sigma^+, v' = 2, 3)^{181}, \qquad O_2(^3\Sigma_g^-, v = 6)^{182}$$

$$S_2(B^2\Sigma_u^-, v' = 8)^{183}, \qquad N_2(B^3\pi_g, v' = 11)^{184}$$

and

$$HCl(v' = 6)$$

as well as the partially discussed $I_2(B^2\pi_{ou+}, v' = 15, 25)^{77-80}$.

When a polyatomic molecule undergoes an electronic transition, dissociation may occur after redistribution of optically excited vibrational energy to non-totally symmetric modes. The probability of dissociation should therefore depend on the total vibrational energy. Crossover to another state can also be induced by the vibrational excitation. Therefore the probability of fluorescence should be increased by a non-quenching gas which removes vibrational energy. Such an effect, first reported by

Terenin and coworkers[185], provides a method for the study of vibrational energy exchange.

The scheme is

(1) $\qquad$ $M + h\nu \quad \rightarrow M^{*\prime}$

(2) $\qquad$ $M^{*\prime} \qquad \rightarrow M$ or products

(3) $\qquad$ $M^{*\prime} + X \rightarrow M^* + X$

(4) $\qquad$ $M^{*\prime} \qquad \rightarrow M + h\nu'$

$\qquad$ $M^* \qquad \rightarrow M + h\nu''$

k_4 may, as an approximation, be assumed to be independent of the amount of vibrational energy. If F_0 is the fluorescence intensity in the absence of added gas and F_X that at added gas concentration $[X]$

$$\frac{F_X}{F_0} = 1 + \frac{k_2 k_3 [X]}{k_4 (k_2 + k_4 + k_3 [X])}$$

which can be rearranged to

$$\frac{F_0}{F_X - F_0} = \frac{k_4}{k_2} + \left(\frac{k_4}{k_2 + k_1}\right)(k_2 k_3 [X])^{-1}$$

Graphical treatment allows

$$\frac{k_3}{k_2 + k_4} = k_3 \tau$$

to be obtained, where τ is the lifetime of the excited molecule. A typical plot of data of this type is that for aniline shown in Figure 1.9.

In the case of aniline, for example, oxygen quenches the fluorescence even at very low pressure and produces no redistribution of the band intensities. It can be assumed to have unit quenching efficiency, and τ determined. k_3 can then be calculated for different foreign gases. The least satisfactory part of the scheme is the assumption that k_4 is independent of the vibrational energy of the molecule. Curme and Rollefson[186] found that the measured quenching constant of β-naphthylamine fluorescence by carbon tetrachloride increases with inert gas pressure owing to an increase in following loss of vibrational energy. Neporent[187] analysed data for naphthylamine and derived the lifetimes of the excited molecules under different conditions (Table 1.17).

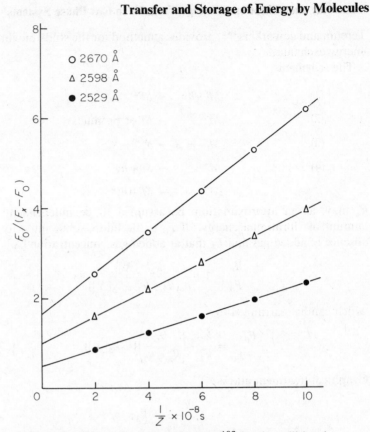

Figure 1.9   Plot of the data of Neporent[187] for the collisional enhancement of fluorescence of aniline. Z is the collision number, F_0 and F_X are the fluorescence intensities in the absence and presence of added gas

A more sophisticated kinetic treatment of the situation has been made by Boudart and Dubois[188], who analyse the implications of multistage collision stabilization through a sequence of vibrational states

$$M_1^{*1}, M_2^{*1}, M_3^{*1}, \dots M_n^*$$

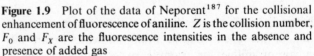

$$(\tau_1) < (\tau_2) < \dots < (\tau)_n = \tau_0$$

The radiationless transition is considered as a first order process requiring an accumulation of energy in a critical degree of freedom and a rate

Table 1.17 [a]

Lifetime ($\times 10^9$ s) of β-naphthylamine molecules excited under different conditions

Temp. °C	λ_{ex} (Å)						
	3660	3341	3129	3022	2804	2652	2537
130	16·2	15·1	13·0	11·9	8·15	3·85	1·95
151	16·2	14·1	12·4	11·2	6·85	3·55	1·65
172	16·2	13·3	11·2	10·4	6·00	2·65	1·40
193	16·2	12·8	10·8	10·0	4·95	2·30	1·05

[a] Data from Neporent[187].

constant written as

$$k = v \exp\left(-E/RT_{vib}\right)$$

T_{vib} is a quantity which depends upon the excess of vibrational energy E in the electronically excited state. If the energy is distributed rapidly among the modes of vibration in the molecule for a gas at temperature T

$$T_{vib} = T + \Delta E/C_{vib}$$

Transfer of energy from an excited molecule by collision will reduce T_{vib} and therefore k.

The coefficient of accommodation α characterizes the efficiency of a gas in removing vibrational energy and provides a measure of the probability of energy transfer during a molecular collision. In a collision between an excited molecule M (vibrational temperature $T_{1\,vib}$) and a foreign molecule X (temperature T_1) an amount of energy E may be transferred to X. The vibrational temperature of M will drop to $T_{2\,vib}$ and its fluorescent lifetime will increase while the temperature of X will increase to T_2.

$$\alpha = \frac{T_2 - T_1}{T_{2\,vib} - T_1}$$

If there is no transfer $T_2 = T_1$, i.e. $\alpha = 0$. For complete equilibrium $\alpha = 1$ and $T_2 = T_{2\,vib}$.

To obtain α it is necessary to evaluate the excess energy as a function of wavelength and temperature. In the case of β-naphthylamine this was done by measuring the fluorescence quenching by oxygen.

Most of the work on this topic has been done with aniline[185], β-naphthyl-amine[187,188,189] and amino-phthalimides[190,191,192]. Typical data is shown for β-naphthylamine in Table 1.18. ΔE increases with the vibrational excitation $(hv_{ex} - hv_0)$ and the molecular complexity (or heat capacity of the additive).

Stevens[189] has made an estimate of collision duration from the transfer data for β-naphthylamine, and finds values increasing from 6 to 14×10^{-13} s in going from methane to n-butane. This leads him to suggest 'wrestling' collisions are important in the transfer of energy between large molecules[2]. Fluorescence studies indicate that up to several kcal/mole can be transferred by collision with foreign gases, the amount of energy transferred increasing with the molecular weight and polarity of the gas[193].

This type of excited state stabilization is very important in gas phase photochemistry. Collisional deactivation can reduce decomposition of the excited states produced by direct light absorption or intersystem crossing. Hexafluoracetone provides an example of the effect of both enhancement of fluorescence[194] and also quenching of decomposition[195]. The system is more complex than originally thought[196] but the observations can be described approximately in the following way.

$$CF_3COCF_3 + hv \rightarrow {}^1CF_3COCF^*_3$$

$${}^1CF_3COCF^*_3 \rightarrow 2CF_3 + CO$$

$${}^1CF_3COCF^*_3 + M \rightarrow {}^1CF_3COCF_3$$

$${}^1CF_3COCF_3 \rightarrow CF_3COCF_3 + hv_f$$

$${}^1CF_3COCF_3 \rightarrow {}^3CF_3COCF^*_3$$

$${}^3CF_3COCF^*_3 \rightarrow 2CF_3 + CO$$

$${}^3CF_3COCF^*_3 + M \rightarrow {}^3CF_3COCF_3 + M$$

$${}^3CF_3COCF_3 \rightarrow CF_3COCF_3$$

$${}^3CF_3COCF_3 \rightarrow CF_3COCF_3 + hv_P$$

There is dispute as to whether deactivation of the upper state, i.e. $CF_3COCF^*_3$, involves a multistep cascade[197] through a number of vibrational levels.

A very large number of such examples could be quoted and are now being resolved by combined studies of light emissions, product analysis and selective energy transfer effects.

Table 1.18[a]

Average amounts of vibrational energy ΔE removed per collision, and accommodation coefficients α, for stabilization of β-naphthylamine ($\bar{\nu}_0 = 29{,}200$ cm^{-1}) by additive X.

X	$T°C$	$\bar{\nu}_{ex}$(cm^{-1}) = 31,930 / E(cm^{-1}) = 2730 ΔE(cm^{-1})	α	33,090 / 3890 ΔE(cm^{-1})	α	35,660 / 6460 ΔE(cm^{-1})	α	37,660 / 8460 ΔE(cm^{-1})	α	39,400 / 10,200 ΔE(cm^{-1})	α	Ref.
He	150							30	0·04	30	0·04	187
H$_2$	150							30	0·03	50	0·05	187
D$_2$	186							50	0·06			188
N$_2$	150					50	0·07	90	0·10			187
CO$_2$	150	20	0·05	50	0·09	120	0·17	430	0·35	170	0·16	187
NH$_3$	150	20	0·05	50	0·09	190	0·20	750	0·63	630	0·43	187
NH$_3$	190					160	0·18	900	0·77			187
CHCl$_3$	150	100	0·16	150	0·17	300	0·22	950	0·55	1100	0·80	187
SF$_6$	186							570	0·24	1250	0·90	188
CH$_4$	180					70	0·09	310	0·30	1600	0·80	189
C$_2$H$_6$	180					170	0·12	530	0·30			189
C$_3$H$_8$	180					340	0·16	790	0·30			189
n-C$_4$H$_{10}$	180					500	0·18	1130	0·34			187
i-C$_5$H$_{12}$	150	60	0·06	120	0·08	300	0·13	1150	0·37	1950	0·52	187
i-C$_5$H$_{12}$	190	50	0·05	100	0·07	330	0·14	1350	0·44	1900	0·47	189
neo-C$_5$H$_{12}$	180					760	0·23	1370	0·33			189
n-C$_6$H$_{14}$	180					1030	0·28	1750	0·38			189
C$_6$H$_6$	150					300	0·17	1150	0·52	3000	1·15	187

[a] Table of data presented by Stevens[2].

1.5 OTHER ENERGY TRANSFER SYSTEMS

In this article only energy transfer effects in electronically excited atoms and molecules have been discussed. Vibrational energy transfer can be studied by a number of methods which have not so far been mentioned. These are sound dispersion, shock wave studies, flash spectroscopy, chemical activation and stabilization and low pressure unimolecular reaction systems. These cannot be described without considerably extending the scope of the article. A few remarks are perhaps in order about energy transfer to or from higher vibrational levels in view of their significance in understanding the behaviour of excited states.

(1) Collisional stabilization of practically monoenergetic ethyl and sec-butyl radicals (H + butene) has been very extensively studied by Rabinovitch[198]. A molecule such as butene-2 may remove energy of the order of 9 or more kcal/mole in a single (strong) collision. For monatomic gases, multistep deactivation occurs with calculated step sizes of 2 to 3 or 1 to 2 kcal/mole (depending on the theoretical model assumed). Studies on vibrationally excited cyclopropane, methyl and dimethyl cyclopropanes[199] (produced by reaction of CH_2 with an olefin), in which vibrational excitation can be very large, lose energy in units of more than 12–15 kcal/mole by collision with polyatomic molecules.

(2) The study of energy transfer by comparing the relative collisional transition probabilities in the low pressure near second order region of a thermal unimolecular reaction is not in a satisfactory state. Johnston[200] has pointed out that studies on the effect of additive gases are only straightforward in the second order region. In practice an empirical comparison is made with the reactant itself[201]. The most recent study which has been made is that of Fletcher, Rabinovitch, Watkins, and Locker[202]. Their data are given in Table 1.19. These authors critically discuss the correlations which may be made with the physical parameters of the activating molecules. They show that if the collisional transition probability for

$$M + X \rightarrow M^* + X$$

to show a direct relationship to parameters such as boiling point[203] is the deactivation probability for

$$M^* + X \rightarrow M + X$$

should be less than unity. This is contrary to what has commonly been assumed. Clearly much revolves around whether a molecule deactivates in a strong or weak collision.

Table 1.19 [a]

Measured collision efficiencies of various gases in energy transfer in the thermal isomerization of methyl isocyanide

Gas	Relative activation efficiency (pressure for pressure basis)	Relative activation efficiency (collision for collision)
CH_3NC	(1·00)	(1·00)
He	0·125	0·14
Ne	0·094	0·16
Ar	0·110	0·17
Kr	0·109	0·18
Xe	0·099	0·15
H_2	0·244	0·15
N_2	0·171	0·24
CO_2	0·462	0·60
CH_4	0·341	0·35
CD_3F	0·889	0·46
CHF_3	0·421	0·52
CF_4	0·289	0·38
C_2H_6		~0·6
HCN	0·527	0·53
C_2H_5CN	0·703	0·66
$n\text{-}C_3H_7CN$	0·937	0·84
CF_3CN	0·440	0·51
CH_3CCH	0·484	0·51
C_2H_5CCH	0·685	0·66

[a] Data of Fletcher and coworkers[202].

(3) The overall process of atomic recombination represented by

$$2A + X \rightarrow A_2 + X$$

where X is a chemically inert third body or 'chaperon' has been very extensively investigated by flash photolysis[204] and other methods. Porter[205] has discussed the problem in terms of two-step two-body collision mechanisms. The most recent data is given in Table 1.20.

The rate constants exhibit a good correlation with the boiling point (Russell and Simons)[204] of the added chaperon and decrease with temperature.

Similar efficiencies can be measured for the recombination of simple radicals.

Table 1.20[a]

Rate constants for the recombination of iodine
atoms in the presence of different additives
at 300°K

Additive	$k(\times 10^{-9}) M^{-2} s^{-1}$
He	1·5, 1·4
Ar	3·0, 3·0, 2·9
H_2	5·7
O_2	6·8
CO_2	13·4
C_4H_{10}	36
C_6H_6	80, 105
CH_3I	160
$C_6H_5CH_3$	194
C_2H_5I	262
$C_6H_3(CH_3)_3$	405
I_2	1600

[a] Data given by Porter[205].

Radiative emission can be responsible for stabilization of the recombination complex of simple molecules and atoms. An example is provided by the well-known 'air-afterglow' reaction:

$$NO + O \rightarrow NO*_2 \rightarrow NO_2 + h\nu$$

REFERENCES

1. P. Pringsheim, *Fluorescence and Phosphorescence*, Interscience, New York (1949).
2. J. C. McCoubrey and W. D. McGrath, *Quart. Rev.* (*London*), **11**, 87 (1957); K. F. Herzfeld and T. A. Litovitz, *Absorption and Dispersion of Ultrasonic Waves*, Academic Press, New York, 1959; T. L. Cottrell and J. C. McCoubrey, *Molecular Energy Transfer in Gases*, Butterworth, London, 1961; B. Stevens, *Collisional Activation in Gases*, Pergamon Press, Oxford, 1967; J. D. Lambert, *Quart. Rev.* (*London*), **21**, 67 (1967).
3. E. W. R. Steacie, *Atomic and Free Radical Reactions*, Reinhold, New York, 1954.
4. R. J. Cvetanović in *Progress in Reaction Kinetics*, Vol. 2 (Ed. G. Porter), Pergamon Press, Oxford, 1964.
5. K. J. Laidler, *The Chemical Kinetics of Excited States*, Oxford University Press, 1955.

6. A. B. Callear, *Appl. Opt. (Suppl. 2)*, **1965**, 145.
7. A. B. Callear and R. J. Oldman, *Trans. Faraday Soc.*, **64**, 840 (1968).
8. A. B. Callear in *Photochemistry and Reaction Kinetics* (Ed. F. S. Dainton, P. G. Ashmore and T. M. Sugden), Cambridge University Press, 1967.
9. A. C. G. Mitchell and M. W. Zemansky, *Resonance Radiation and Excited Atoms*, Cambridge University Press, 1934.
10. M. W. Zemansky, *Phys. Rev.*, **36**, 919 (1930).
11. E. A. Milne, *J. Math. Soc. (London)*. **1**, 1 (1926) and earlier paper; also L. M. Bieberman, *J. Expt. Theor. Phys. (USSR)*, **9**, 584 (1949).
12. E. W. Samson, *Phys. Rev.*, **40**, 940 (1932).
13. C. G. Matland, *Phys. Rev.*, **92**, 637 (1953).
14. T. Holstein, *Phys. Rev.*, **72**, 1212 (1947); *Phys. Rev.*, **83**, 1159 (1951).
15. A. J. Yarwood, O. P. Strausz and H. E. Gunning, *J. Chem. Phys.*, **41**, 1705 (1964).
16. R. J. Cvetanović, *J. Chem. Phys.*, **23**, 1203, 1208 (1955).
17. R. B. Cundall and T. F. Palmer, *Trans. Faraday Soc.*, **56**, 1211 (1960).
18. B. de B. Darwent and F. G. Hurtubise, *J. Chem. Phys.*, **20**, 1684 (1952).
19. A. B. Callear and R. G. W. Norrish, *Proc. Roy. Soc. (London)*, **A266**, 299 (1962).
20. A. B. Callear and G. J. Williams, *Trans. Faraday Soc.*, **60**, 2158 (1964).
21. M. D. Scheer and J. Fine, *J. Chem. Phys.*, **36**, 1264 (1962).
22. S. Penzes, A. J. Yarwood, O. P. Strausz and H. E. Gunning, *J. Chem. Phys.*, **43**, 4524 (1965).
23. J. A. Berberet and K. C. Clark, *Phys. Rev.*, **100**, 506 (1955).
24. G. H. Kimbell and D. J. Le Roy, *Can. J. Chem.*, **40**, 1229 (1962); see also J. E. McAlduff and D. J. Le Roy, *Can. J. Chem.*, **43**, 2279 (1965).
25. S. Mrozowski, *Rev. Mod. Phys.*, **16**, 153 (1944).
26. A. B. Callear and W. J. R. Tyerman, *Nature*, **202**, 1326 (1964); *Trans. Faraday Soc.*, **62**, 2313 (1966).
27. J. D. Lambert and R. Salter, *Proc. Roy. Soc. (London)*, **A253**, 277 (1957).
28. A. B. Callear and R. J. Oldman, *Trans. Faraday Soc.*, **63**, 2888 (1967).
29. A. B. Callear and R. J. Oldman, *Trans. Faraday Soc.*, **64**, 840 (1968).
30. R. J. Donovan and D. Husain, *Trans. Faraday Soc.*, **62**, 11, 1050 (1966).
31. H. Yamazaki and R. J. Cvetanović, *J. Chem. Phys.*, **41**, 3703 (1964).
32. K. F. Preston and R. J. Cvetanović, *J. Chem. Phys.*, **45**, 2888 (1966).
33. D. R. Bates, *Proc. Phys. Soc.*, **73**, 227 (1959).
34. W. R. Thorson, *J. Chem. Phys.*, **34**, 1744 (1961).
35. W. R. Thorson and J. W. Moskowitz, *J. Chem. Phys.*, **38**, 1848 (1963).
36. E. E. Nikitin, *J. Chem. Phys.*, **43**, 744 (1965).
37. V. K. Bichovskii and E. E. Nikitin, *Opt. Spectr. (USSR) (English Transl.)*, **16**, 111 (1964).
38. R. G. W. Norrish and W. MacF. Smith, *Proc. Roy. Soc.*, **A176**, 295 (1940); W. MacF. Smith and F. W. Southam, *J. Chem. Phys.*, **15**, 845 (1947).
39. P. G. Dickens, J. W. Linnett and O. Sovers, *Discussions Faraday Soc.*, **33**, 52 (1962).
40. See references 1 and 5 and A. Jablonski, *Z. Physik*, **70**, 723 (1931).

41. J. L. Magee, *J. Chem. Phys.*, **8**, 687 (1940); J. L. Magee and T. Ri, *J. Chem. Phys.*, **9**, 638 (1941).
42. K. J. Laidler, *J. Chem. Phys.*, **10**, 34, 43 (1942); K. J. Laidler, *J. Chem. Phys.*, **15**, 712 (1947); J. C. Polanyi, *J. Quant. Spect. and Radiative Transfer*, **3**, 471 (1963).
43. E. Wigner, *Gottingen Nachrichten*, **1927**, 375.
44. G. Karl and J. C. Polanyi, *Discussions Faraday Soc.*, **33**, 93 (1962); *J. Chem. Phys.*, **38**, 271 (1963).
45. G. Karl, P. Kruus, and J. C. Polanyi, *J. Chem. Phys.*, **46**, 224 (1967).
46. G. Karl, P. Kruus and J. C. Polanyi, *J. Chem. Phys.*, **46**, 244 (1967).
47. J. B. Homer and F. P. Lossing, *Can. J. Chem.*, **44**, 143 (1966).
48. H. E. Gunning and O. P. Strausz in *Advances in Photochemistry*, Vol. 1 (Ed. W. A. Noyes, Jr., G. S. Hammond and J. N. Pitts, Jr.), Wiley, New York, 1963.
49. R. J. Cvetanović, *Can. J. Chem.*, **38**, 1678 (1960).
50. O. P. Strausz and H. E. Gunning, *Advances in Photochemistry*, Vol. 4 (Ed. W. A. Noyes, Jr., G. S. Hammond and J. N. Pitts, Jr.), Wiley, New York, 1966.
51. O. P. Strausz and H. E. Gunning, *J. Am. Chem. Soc.*, **84**, 4080 (1962).
52. A. R. Trobridge and K. R. Jennings, *Proc. Chem. Soc.*, **1964**, 335.
53. S. Penzes, O. P. Strausz and H. E. Gunning, *J. Chem. Phys.*, **45**, 2322 (1966).
54. F. P. Lossing and T. F. Palmer, *Can. J. Chem.*, **41**, 2412 (1963).
55. K. Yang, *J. Am. Chem. Soc.*, **86**, 3941 (1964); *J. Am. Chem. Soc.*, **87**, 5294 (1965); *J. Am. Chem. Soc.*, **88**, 4575 (1966); *J. Am. Chem. Soc.*, **89**, 5344 (1967).
56. A. G. Gaydon and I. R. Hurle, *The Shock Tube in High Temperature Chemistry and Physics*, Chapman & Hall, London, 1963; I. R. Hurle, *J. Chem. Phys.*, **41**, 3911 (1964).
57. W. L. Starr, *J. Chem. Phys.*, **43**, 73 (1965).
58. J. E. Mentall, H. F. Krause and W. L. Fite, *Discussions Faraday Soc.*, **44**, 157 (1967).
59. D. R. Jenkins, *Proc. Roy. Soc.*, **A293**, 493 (1966).
60. J. Franck, *Z. Physik*, **9**, 259 (1922).
61. G. Cario and J. Franck, *Z. Physik*, **11**, 161 (1922).
62. H. S. W. Massey and E. H. S. Burhop, *Electronic and Ionic Impact Phenomena*, Oxford University Press, 1952.
63. N. F. Mott and H. S. W. Massey, *The Theory of Atomic Collisions*, Oxford University Press, 1949.
64. B. Beutler and H. Josephy, *Z. Physik*, **53**, 747 (1929).
65. R. A. Anderson and R. H. McFarland, *Phys. Rev.*, **119**, 693 (1960).
66. E. E. Stepp and R. A. Anderson, *J. Opt. Soc. Am.*, **55**, 31 (1965).
67. A. Javan, W. R. Bennett, Jr. and D. R. Herriott, *Phys. Rev. Letters*, **6**, 106 (1961).
68. E. E. Benton, F. A. Matson, E. E. Ferguson and W. W. Roberts, *Phys. Rev.*, **128**, 206 (1962).
69. D. J. Le Roy and E. W. R. Steacie, *J. Chem. Phys.*, **9**, 829 (1941); R. J. Cvetanović, H. E. Gunning and E. W. R. Steacie, *J. Chem. Phys.*, **31**, 573 (1959).

70. J. R. McNesby and H. Okabe in *Advances in Photochemistry* (Ed. W. A. Noyes, Jr., G. S. Hammond, J. N. Pitts, Jr.), Wiley, New York, 1965.
71. G. von Bunau, *Fortschr. Chem. Forsch.*, **5**, 347 (1965).
72. W. Groth (1954), *Z. Physik Chem. (Frankfurt)*, **NF1**, 300 (1954); W. Groth, *Z. Elektrochem. (Frankfurt)*, **58**, 752 (1954); W. Groth, W. Pessara and H. J. Rommel, *Z. Phys. Chem.*, **NF32**, 192 (1962).
73. J. R. McNesby and H. Okabe in *Advances in Photochemistry* (Ed. W. A. Noyes, Jr., G. S. Hammond, J. N. Pitts, Jr.), Wiley, New York, 1965.
74. G. von Bunau and R. N. Schindler, *J. Chem. Phys.*, **44**, 420 (1966); *Munich Conference on Photochemistry*, Preprints (1967).
75. W. R. Bennett, Jr., W. L. Faust, R. A. McFarlane and C. K. M. Patel, *Phys. Rev. Letters*, **8**, 470 (1962).
76. I. Tanaka and E. W. R. Steacie, *Can. J. Chem.*, **35**, 821 (1957); I. Tanaka and E. W. R. Steacie, *J. Chem. Phys.*, **26**, 715 (1957); I. Tanaka and J. R. McNesby, *J. Chem. Phys.*, **36**, 3170 (1962); M. Yoshida and I. Tanaka, *J. Chem. Phys.* **44**, 494 (1966).
77. C. Arnot and C. A. McDowell, *Can. J. Chem.*, **36**, 114 (1958).
78. B. Stevens, *Can. J. Chem.*, **36**, 831 (1959).
79. J. C. Polanyi, *Can. J. Chem.*, **36**, 121 (1958).
80. R. L. Brown and W. Klemperer, *J. Chem. Phys.*, **41**, 3072 (1964); J. I. Steinfeld and W. Klemperer, *J. Chem. Phys.*, **42**, 3475 (1965).
81. H. J. Bauer, H. O. Kneser and E. Sittig, *J. Chem. Phys.*, **30**, 1119 (1959).
82. B. Brocklehurst and K. R. Jennings in *Progress in Reaction Kinetics*, Vol. 4, (Ed. G. Porter), Pergamon Press, Oxford, 1967.
83. A. B. Callear and I. W. M. Smith, *Trans. Faraday Soc.*, **61**, 2383 (1965).
84. N. H. Sagert and B. A. Thrush, *Discussions Faraday Soc.*, **37**, 223 (1964).
85. W. R. Brennen and G. B. Kistiakowsky, *J. Chem. Phys.*, **44**, 2695 (1966).
86. Lord Rayleigh, *Proc. Roy. Soc.*, **A85**, 219 (1911).
87. C. G. Freeman and L. F. Phillips, *J. Phys. Chem.*, **68**, 362 (1964).
88. L. F. Phillips, *Can. J. Chem.*, **43**, 369 (1965).
89. R. A. Young, *Can. J. Chem.*, **43**, 3228 (1965).
90. F. Kaufman and J. R. Kelso, *Discussions Faraday Soc.*, **37**, 26 (1964); A. Mathias and H. I. Schiff, *J. Chem. Phys.*, **10**, 3118 (1964).
91. L. W. Bader and E. A. Ogryzlo, *Discussions Faraday Soc.*, **37**, 46 (1964).
92. e.g. J. T. Vanderslice, E. A. Mason and W. G. Maisch, *J. Chem. Phys.*, **31**, 738 (1959); O. P. Stausz and H. E. Gunning, *Can. J. Chem.*, **39**, 2549 (1961).
93. A. Terenin and A. V. Karyakin, *Izv. Akad. Nauk SSSR Ser. Fiz.*, **15**, 550 (1951); *Dokl. Akad. Nauk SSSR*, **96**, 269 (1954).
94. F. S. Dainton and K. J. Ivin, *Trans. Faraday Soc.*, **46**, 374, 382 (1950).
95. R. B. Cundall, *Progress in Reaction Kinetics*, Vol. 2 (Ed. G. Porter), Pergamon Press, Oxford, 1964.
96. K. Greenhough and A. B. F. Duncan, *J. Am. Chem. Soc.*, **83** (1961).
97. J. T. Dubois and W. A. Noyes, Jr., *J. Chem. Phys.*, **19**, 1512 (1951).
98. R. B. Cundall, F. J. Fletcher and D. G. Milne, *J. Chem. Phys.*, **39**, 3536 (1963); *Trans. Faraday Soc.*, **60**, 1146 (1964).

99. S. Sato, K. Kikuchi and M. Tanaka, *J. Chem. Phys.*, **39**, 239 (1963); M. Tanaka, T. Terao and S. Sato, *Bull. Chem. Soc. Japan*, **38**, 1645 (1965); M. Tanaka, T. Terumi and S. Sato, *Bull. Chem. Soc. Japan*, **39**, 1423 (1966).

100. J. T. Dubois, *J. Phys. Chem.*, **63**, 638 (1959).

101. B. Stevens, *Discussions Faraday Soc.*, **27**, 34 (1959).

102. B. Stevens, *Chem. Rev.*, **57**, 439 (1957).

103. G. P. Semeluk and I. Unger, *Nature*, **198**, 853 (1963); A. K. Basak, G. P. Semeluk and I. Unger, *J. Phys. Chem.*, **70**, 1337 (1966).

104. e.g. W. A. Noyes, Jr., and I. Unger, *Organic Photochemistry*, I.U.P.A.C. (Strasbourg), Butterworths, London, 1965, p. 1; W. A. Noyes, Jr., and I. Unger in *Advances in Photochemistry*, Vol. 4 (Ed. W. A. Noyes, Jr., G. S. Hammond and J. N. Pitts, Jr.), Wiley, New York, 1966.

105. J. E. Wilson and W. A. Noyes, Jr., *J. Am. Chem. Soc.*, **63**, 3025 (1941).

106. G. B. Kistiakowsky and M. Nelles, *Phys. Rev.*, **41**, 595 (1932); C. K. Ingold and C. L. Wilson, *J. Chem. Soc.*, **1936**, 941; C. L. Wilson, *J. Chem. Soc.*, **1936**, 1210.

107. H. Ishikawa and W. A. Noyes, Jr., *J. Am. Chem. Soc.*, **84**, 1502 (1962); *J. Chem. Phys.* **37**, 583 (1962).

108. J. W. Sidman and D. S. McClure, *J. Am. Chem. Soc.*, **77**, 6461 (1955).

109. J. A. Poole, *J. Phys. Chem.*, **69**, 1343 (1965).

110. R. B. Cundall and A. S. Davies, *Trans. Faraday Soc.*, **62**, 1151 (1966).

111. W. A. Noyes, Jr., W. A. Mulac and D. A. Harter, *J. Chem. Phys.*, **44**, 2100 (1966).

112. G. M. Almy and P. R. Gillette, *J. Chem. Phys.*, **11**, 188 (1943).

113. W. A. Noyes, Jr. and D. A. Harter, *J. Chem. Phys.*, **46**, 674 (1967).

114a G. A. Haninger, Jr. and E. K. C. Lee, *J. Phys. Chem.*, **71**, 3104 (1967).

114b R. B. Cundall and K. Dunnicliff, to be published.

115. P. Sigal, *J. Chem. Phys.*, **42**, 1953 (1965); *J. Chem. Phys.*, **46**, 1043 (1967).

116. G. B. Kistiakowsky and C. S. Parmenter, *J. Chem. Phys.*, **42**, 2942 (1965).

117. S. J. Strickler and R. J. Watts, *J. Chem. Phys.*, **44**, 426 (1966).

118. See discussion in *The Triplet State* (Proceedings of Conference in Beirut), Cambridge University Press, 1967.

119. R. B. Cundall, A. S. Davies and K. Dunnicliff in *The Triplet State* (Proceedings of Conference in Beirut), Cambridge University Press, 1967, p. 183.

120. C. S. Parmenter and B. L. Ring, *J. Chem. Phys.*, **46**, 1998 (1967).

121. D. S. McClure, *J. Chem. Phys.*, **17**, 905 (1949), and also M. R. Wright, R. P. Frosch and G. W. Robinson, *J. Chem. Phys.*, **33**, 936 (1960).

122. C. L. Braun, S. Kato and S. Lipsky, *J. Chem. Phys.*, **39**, 1645 (1963).

123. J. K. Foote, M. H. Mallon and J. R. Pitts, *J. Am. Chem. Soc.*, **88**, 3698 (1966).

124. K. Shindo and S. Lipsky, *J. Chem. Phys.*, **45**, 2292 (1966).

125. H. R. Ward, J. S. Wishnok, P. D. Sherman, *J. Am. Chem. Soc.*, **89**, 162 (1967).

126. L. Kaplan and K. E. Wilzbach, *J. Am. Chem. Soc.*, **89**, 1030 (1967).

127. K. E. Wilzbach, J. S. Ritscher and L. Kaplan, *J. Am. Chem. Soc.*, **89**, 1031 (1967); *J. Am. Chem. Soc.*, **90**, 1116 (1968).

128. K. E. Wilzbach, A. L. Harkness and L. Kaplan, *Proceedings of Munich Photochemistry Meeting*, 1967.

129. W. M. Jackson, J. L. Faris and B. Donn, *J. Phys. Chem.*, **71**, 3346 (1967).

130. R. R. Hentz and S. J. Rzad, *J. Phys. Chem.*, **71**, 4096 (1967).
131. R. B. Cundall and K. Dunnicliff, in course of publication.
132. I. Unger, *J. Phys. Chem.* **69**, 4884 (1965).
133. D. Phillips, *J. Phys. Chem., ***71**, 1839 (1967).
134. D. Phillips, *J. Chem. Phys.*, **46**, 4679 (1967).
135. I. Haller, *J. Am. Chem. Soc.*, **88**, 2070 (1966); *J. Chem. Phys.*, **47**, 1117 (1967).
136. J. Lemaire, *J. Phys. Chem.*, **71**, 612 (1967).
137. H. R. Ward, *J. Am. Chem. Soc., ***89**, 2367 (1967).
138. L. Kaplan, K. E. Wilzbach, W. G. Brown and S. S. Yang, *J. Am. Chem. Soc.*, **87**, 675 (1965).
139. S. Hirokami and S. Sato, *Can. J. Chem.*, **45**, 3181 (1967); T. Terao, S. I. Hirokami, S. Sato and R. J. Cvetanović, *Can. J. Chem.*, **44**, 2173 (1966).
140. L. S. Kassel, *The Kinetics of Homogeneous Gas Reaction*, Chemical Catalog Co., New York, 1932.
141. R. B. Cundall and F. J. Fletcher, unpublished.
142a. R. J. Cvetanović, *Can. J. Chem.*, **38**, 1678 (1960).
142b R. J. Cvetanović in *Advances in Photochemistry*, Vol. 1 (Ed. W. A. Noyes, Jr., G. S. Hammond, J. N. Pitts, Jr.), Wiley, New York, 1963.
143. D. H. Shaw and H. O. Pritchard, *J. Phys. Chem., ***70**, 1230 (1966).
144. R. E. Rebbert and P. Ausloos, *J. Am. Chem. Soc., ***89**, 1573 (1967).
145. E. K. C. Lee and N. W. Lee, *J. Phys. Chem.*, **71**, 1167 (1967).
146. E. K. C. Lee, *J. Phys. Chem.*, **71**, 2804 (1967).
147. S. Y. Ho and W. A. Noyes, Jr., *J. Am. Chem. Soc., ***89**, 5091 (1967).
148. R. B. Cundall and A. S. Davies, *Trans. Faraday Soc.*, **62**, 2444 (1966).
149. R. B. Cundall and A. S. Davies in *Progress in Reaction Kinetics*, Vol. 4 (Ed. G. Porter), Pergamon Press, Oxford, 1967.
150. R. B. Cundall and A. S. Davies, *Proc. Roy. Soc.*, **A290**, 563 (1966).
151. R. B. Cundall and A. S. Davies, *Trans. Faraday Soc.*, **62**, 2793 (1966).
152. G. W. Robinson, *J. Mol. Spectry.*, **6**, 58 (1961).
153. G. W. Robinson and R. P. Frosch, *J. Chem. Phys.*, **37** (1962).
154. G. W. Robinson and R. P. Frosch, *J. Chem. Phys.*, **38**, 1187 (1962).
155. M. Gouterman, *J. Chem. Phys.*, **36**, 2846 (1962); see also P. Seybold and M. Gouterman, *Chem. Rev., ***65**, 413 (1965) and S. K. Lower and M. A. El-Sayed, *Chem. Rev.*, **66**, 200 (1966).
156. E. F. McCoy and I. G. Ross, *Australian J. Chem.*, **15**, 573 (1962); G. R. Hunt, E. F. McCoy and I. G. Ross, *Australian J. Chem.*, **15**, 591 (1962); J. P. Bryne, E. F. McCoy and I. G. Ross, *Australian J. Chemistry*, **18**, 1589 (1965).
157. G. W. Robinson in *The Triplet State* (Proceedings of Conference in Beirut), Cambridge University Press, 1967; *J. Chem. Phys., ***47**, 1967 (1967).
158. S. D. Colson and E. R. Bernstein, *J. Chem. Phys.*, **43**, 2661 (1965).
159. E. C. Lim, *J. Chem. Phys.*, **36**, 3497 (1962).
160. D. Bryce-Smith and H. C. Longuet-Higgins, *Chem. Comm.*, **1966**, 593.
161. W. Siebrand, *J. Chem. Phys., ***44**, 4055 (1966); *J. Chem. Phys.*, **46**, 440 (1967).
161a W. Siebrand, in *The Triplet State* (Proceedings of Conference in Beirut), Cambridge University Press, 1967.
162. M. R. Wright, R. P. Frosch and G. W. Robinson, *J. Chem. Phys.*, **33**, 934 (1960).

163. P. D. Dikun, *Zh. Eksperim. i Teor. Fiz.*, **20**, 193 (1950).
164. R. Williams, *J. Chem. Phys.*, **28**, 577 (1958).
165. B. Stevens and P. C. McCartin, *Mol. Phys.*, **3**, 425 (1960); B. Stevens, E. Hutton and G. Porter, *Nature*, **186**, 1045 (1960).
166. C. A. Parker and C. G. Hatchard, *Proc. Chem. Soc.*, **1962**, 147; *Proc. Roy. Soc.*, **A269**, 574 (1962).
167. G. Porter and F. J. Wright, *Trans. Faraday Soc.*, **51**, 1205 (1955).
168. G. Porter and P. West, *Proc. Roy. Soc.*, **A279**, 302 (1964).
169. B. Stevens, M. S. Walker and E. Hutton, in *The Triplet State* (Proceedings of Conference in Beirut), Cambridge University Press, 1967.
170. B. Stevens and M. S. Walker, *Chem. Comm.*, **1965**, 18.
171. A. B. Zahlan, S. Z. Weisz, R. C. Jarnigan and M. Silver, *J. Chem. Phys.*, **42**, 4244 (1965).
172. G. Finger, O. Zamoni-Khamini, J. Olmsted and A. B. Zahlan in *The Triplet State* (Proceedings of Conference in Beirut), Cambridge University Press, 1967.
173. J. Perrin, *2me Conseil de Chimie, Solway*, Gauthier-Villars, Paris, 1925.
174. T. Förster, *Z. Elektrochem.*, **56**, 716 (1952) and later papers.
175. D. L. Dexter, *J. Chem. Phys.*, **21**, 836 (1953).
176. E. J. Bowen and B. Brocklehurst, *Trans. Faraday Soc.*, **49**, 1131 (1953).
177. P. Pringsheim, *Ann. Acad. Sci. Tech. Varsovie*, **5**, 29 (1938).
178. B. Stevens and E. Hutton, *Mol. Phys.*, **3**, 71 (1960).
179. R. Williams and G. J. Goldsmith, *J. Chem. Phys.*, **39**, 2008 (1963).
180. R. J. Watts and S. J. Strickler, *J. Chem. Phys.*, **44**, 2423 (1966).
181. A. B. Callear and I. W. M. Smith, *Trans. Faraday Soc.*, **59**, 1735 (1963).
182. F. J. Lipscomb, R. G. W. Norrish and B. A. Thrush, *Proc. Roy. Soc.*, **A233**, 455 (1956).
183. E. Durand, *J. Chem. Phys.*, **8**, 46 (1940).
184. C. R. Stanley, *Proc. Roy. Soc.*, **A241**, 180 (1957); J. F. Noxon, *J. Chem. Phys.*, **36**, 926 (1962).
185. A. N. Terenin, A. T. Vartanyan and B. S. Neporent, *Trans. Faraday Soc.*, **35**, 39 (1939); B. S. Neporent, *Zh. Fiz. Khim.*, **13**, 965 (1939).
186. H. C. Curme and G. K. Rollefson, *J. Am. Chem. Soc.*, **74**, 28 (1952).
187. B. S. Neporent, *Zh. Fiz. Khim.*, **21**, 1111 (1947); *Zh. Fiz. Khim.*, **24**, 1219 (1950).
188. M. Boudart and J. T. Dubois, *J. Chem. Phys.*, **23**, 223 (1955).
189. B. Stevens, *Mol. Phys.*, **3**, 589 (1960).
190. N. A. Borisevich and B. S. Neporent, *Opt. Spectry. (USSR)*, **1**, 536 (1956).
191. B. S. Neporent and S. O. Mirumyants, *Opt. Spectry. (USSR)*, **8**, 635 (1960).
192. S. O. Mirumyants and B. S. Neporent, *Opt. Spectry. (USSR)*, **8**, 787 (1960).
193. e.g. E. J. Bowen and S. Veljkovic, *Proc. Roy. Soc.*, **A236**, 1 (1956).
194. H. Okabe and E. W. R. Steacie, *Can. J. Chem.*, **36**, 137 (1958).
195. P. Ayscough and E. W. R. Steacie, *Proc. Roy Soc.*, **A234**, 476 (1956).
196. A. Gandini, D. A. Whytock and K. O. Kutschke, *Proc. Munich Conference on Photochemistry*, 1967, and papers quoted there.
197. A. N. Strachan, R. K. Boyd and K. O. Kutschke, *Can. J. Chem.*, **42**, 1345 (1964).

198. e.g. R. E. Harrington, B. S. Rabinovitch and M. R. Hoare, *J. Chem. Phys.*, **33**, 744 (1960); G. H. Kohlmaier and B. S. Rabinovitch, *J. Chem. Phys.*, **38**, 1692, 1709 (1963), and *J. Chem. Phys.*, **39**, 490 (1963).
199. D. W. Setser, B. S. Rabinovitch and J. W. Simons, *J. Chem. Phys.*, **40**, 1751 (1964); **41**, 800 (1965).
200. M. Volpe and H. S. Johnston, *J. Am. Chem. Soc.*, **78**, 3903 (1956).
201. A. F. Trotman-Dickenson, *Gas Kinetics*, Butterworths, London, 1955.
202. F. J. Fletcher, B. S. Rabinovitch, K. W. Watkins and D. J. Locker, *J. Phys. Chem.*, **70**, 2823 (1966).
203. H. S. Johnston, *Ann. Rev. Phys. Chem.*, **8**, 249 (1957).
204. K. E. Russell and J. Simons, *Proc. Roy. Soc.*, **A217**, 271 (1953); M. I. Christie, A. J. Harrison, R. G. W. Norrish and G. Porter, *Proc. Roy. Soc.*, **A231**, 446 (1955); G. Porter and J. A. Smith, *Proc. Roy. Soc.*, **A261**, 28 (1961).
205. G. Porter, *Discussions Faraday Soc.*, **33**, 198 (1962).

2

Thermal Population of Excited Electronic States—Excitation and Emission in Shock Waves

J. N. Bradley

2.1 INTRODUCTION

The study of the efficiency of energy transfer between translational and electronic states requires an experimental technique in which appreciable amounts of energy are fed into one of the states and the subsequent increase of energy in the other states can be observed. Virtually the only methods available at present which meet this requirement are based on shock wave excitation, principally because the aerodynamic properties of the shock transition lead to the direct and rapid conversion (typically within 10^{-9} s) of the flow energy of the gas into translational motions of the molecules. The internal energy modes relax far more slowly, at rates which provide a direct measure of the efficiency of energy transfer. The populations of the excited electronic levels can usually be determined from the intensity of the corresponding emission, although less direct methods can be used if a suitable transition is not available.

In principle, the converse of this experiment is also feasible, that is, excitation of particular electronic levels may be achieved by illuminating the system with radiation of the appropriate wavelength and the subsequent relaxation followed by monitoring the increase in the translational temperature. In practice, this approach cannot yet be used both because high intensity, short duration light sources are available only for a very restricted number of wavelengths and because it is difficult to make an independent measurement of the translational temperature.

For this reason, translational $\leftrightarrow$ electronic energy transfer will be discussed almost entirely in the light of evidence obtained from shock wave experiments. Related studies based, for example, on fluorescence quenching or flame spectroscopy also provide useful information on particular systems which may serve to confirm the shock wave findings.

As will become apparent later, the direct transfer of energy between the two states by simple binary collisions is extremely inefficient, particularly when atomic species only are involved, and energy exchange usually occurs via a multi-stage process. In order to simplify the discussion, the various excitation processes may be classified roughly as follows:

I. *Direct transfer*—translation → electronic excitation

$$T \rightarrow E^*$$

II. *Vibration–electronic energy transfer*

$$T \rightsquigarrow V \rightarrow E^*$$

III. *Electronic energy transfer*—excitation of an electronic state of species *A* followed by exchange with species *B*

$$T \rightsquigarrow E_A^*; \qquad E_A^* + E_B \rightarrow E_A + E_B^*$$

IV. *Chemi-excitation*—initiation of exothermic chemical reactions in which the excess energy appears as electronic excitation

$$T \rightsquigarrow \text{chemical reaction} \rightarrow E^*$$

V. *Excitation by electrons*—excitation by direct interaction with free electrons, following thermal ionization

$$T \rightsquigarrow \text{ionization}, I; \qquad e + E \rightarrow E^* + e$$

VI. *Ion-electron recombination*—excitation by association of ions and electrons, following thermal ionization

$$T \rightsquigarrow I; \qquad A^+ + e \rightarrow E^*$$

This classification should not be considered as exhaustive and various additional or alternative subdivisions could be proposed. As an example, the chemical reaction referred to under Process IV may not produce an excited species but may simply produce a different chemical entity which can effect transfer by Processes I or II so that the chemical influence is catalytic in nature.

The low efficiency of direct transfer may be demonstrated quite readily in terms of simple perturbation theory. A collisional encounter may be considered as a time-dependent perturbation which depends on the nature of the molecular interaction potential and on the dynamics of the collision. The probability of a transition between states m and n is then given by

$$p_{nm} = \frac{4\pi^2}{h^2} \left| \int_{\infty}^{\infty} V_{mn}\, e^{i\omega_{mn}t}\, dt \right|^2$$

where $\omega_{mn} = 2\pi(E_m - E_n)/h$ and V_{mn} is the matrix element of the perturbation between the two states m and n. In this case, the perturbation is a single collision and has a maximum at $t = 0$. The integral will have an appreciable magnitude only if the frequency of the perturbation is close to that of ω_{mn}. Since the frequency associated with a typical electronic transition is about $10^{15}\ \mathrm{s}^{-1}$ and the duration of a gas-phase collision at ambient temperatures is of the order of $10^{-12}\ \mathrm{s}$, direct transfer of energy between translational and electronic states has a very low probability.

For collisional excitation, two potential energy surfaces for the collision complex may be envisaged, one of which contains an atom or molecule in its ground state and the other contains it in an excited state. If a transition is to occur, the two surfaces must intersect, the probability of the transition depending on the angle of intersection and on whether electron spin is conserved during the transition (these two factors primarily constitute the matrix element referred to above). Such surfaces are difficult to compute theoretically, particularly for complex species, and it is therefore not easy to predict a priori whether a transition is likely to occur.

However, if both the collision partners are atomic, the surfaces reduce to simple potential energy curves depending only on a single parameter, the internuclear separation. If one of the atoms is that of an inert gas, the most common situation will correspond to two closely-parallel, non-intersecting curves (Figure 2.1) and, as before, the probability of direct excitation is very low. This prediction has been very clearly demonstrated by fluorescence quenching experiments, where it has been shown that inert gases have a negligible effect in quenching excited mercury or sodium.

Where the forms of the repulsion potential differ markedly between the two states, for example if incipient molecule formation occurs, then crossing becomes more probable and the efficiency of transfer increases (Figure 2.2). For this reason, molecule formation has sometimes been invoked in order to explain the excitation of inert gas atoms, e.g. Xe_2 has been suggested as a precursor for the emission of xenon radiation.

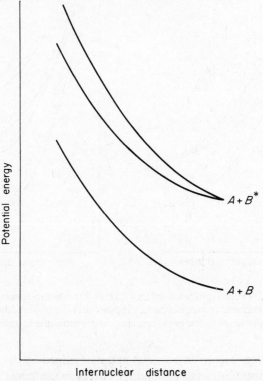

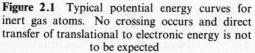

Figure 2.1 Typical potential energy curves for inert gas atoms. No crossing occurs and direct transfer of translational to electronic energy is not to be expected

Clearly, the participation of a molecular species in the collision complex allows for greater variation in the shapes of the potential energy surfaces and therefore enhances the probability of energy transfer. This is a very crude explanation of the greater efficiency of vibrational–electronic energy transfer (Process II), frequently observed, for example, by fluorescence quenching measurements. This process is also promoted by the greater similarity in the frequencies associated with the two transitions, vibrational quanta typically being associated with frequencies one-tenth or more of the corresponding electronic quanta.

The efficiency of energy transfer between electronic states of different species (Process III) depends on the degree of matching of the quantum of

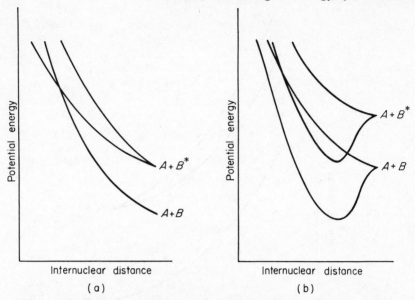

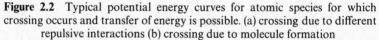

Figure 2.2 Typical potential energy curves for atomic species for which crossing occurs and transfer of energy is possible. (a) crossing due to different repulsive interactions (b) crossing due to molecule formation

electronic energy lost by one species and that gained by the other. When total electron spin is conserved and the energy mismatch is small, a near-resonance condition is achieved and transfer may then become very efficient. This situation may also be discussed on the basis of the perturbation theory approach above, where the subscripts m and n must now each refer to a pair of states, one for each of the two molecules involved. Since the matrix element V_{mn} is taken between two pairs of states it will be reduced in magnitude, but the value of ω_{mn} will be correspondingly reduced and will ultimately vanish at resonance. Since the latter appears in an exponential, the overall effect of the simultaneous change in V_{mn} and ω_{mn} is to make the total integral larger. Cross-sections for near-resonance transfer of this type can come very close to the gas kinetic values.

Probably the most common source of radiation found in shock wave investigations is chemiluminescence, due to the energy released by chemical processes appearing as electronic excitation (Process IV). A wide range of exothermic reactions may be envisaged and the following general types can be distinguished:

Two-body association $\qquad A + B \rightarrow AB^*$ (A)

Three-body association (i) $A + B + C \rightarrow AB^* + C$ (B)

(ii) $A + B + C \rightarrow AB + C^*$ (C)

Exchange reactions (i) $A + BC \rightarrow AB + C^*$ (D)

(ii) $AB + CD \rightarrow AC^* + BD$ (E)

Examples of some of these basic types will be discussed below. Reactions which do not conveniently fall into one of these categories can usually be regarded as a combination of two such processes. Chemiluminescent reactions can, of course, be studied by a wide range of techniques and those based on shock waves are by no means the most informative. Since the shock wave method is the only one which provides genuine thermal excitation as the primary process, information gained by other techniques is normally employed to assist in the interpretation of shock wave findings.

At greater extents of thermal excitation, ionization will take place. The mechanisms responsible for ionization are usually complex and are not fully understood, although direct thermal ionization via binary collisions does participate even in the inert gases. A stepwise process is normally found to occur and so electronically-excited states are involved, being populated by a direct thermal route (Process I). Excited states will also be produced after ionization due to the interaction between neutral and charged species present. The electrons may remain free but transfer energy during an encounter—bremsstrahlung (Process V)—or they may associate, either with neutral species to give negative ions or with ions, by radiative or by three-body recombination. All these processes can populate excited states, the excitation by free electrons being particularly efficient because of the equal masses of the free and the bound electrons, the high velocity of the free electrons, leading to short duration collisions, and the strong Coulombic interaction between the electrons. These mechanisms can naturally only become dominant at very high temperatures ($\sim 10{,}000°\text{K}$) where the degree of ionization becomes significant.

The various mechanisms for excitation of electronic states are described further in the discussion of specific examples taken from shock wave investigations. Because atomic systems are, at least in principle, simpler to interpret, they will be dealt with first.

2.2 EXCITATION OF ATOMIC SPECIES

Emission spectra from a large number of atomic species (e.g. He[1], Ne[2], Kr[3], Ar[4], Xe[3], H[5], Na[6], Rb[7], Cs[8], Mg[7], Ca[6], Sr[7], Ba[6], Ti[7], Cr[9], Fe[6], Cu[6],

Hg^{10}, Al^6) have been observed in the shock tube but in only a very limited number of instances has the kinetic investigation, essential for an identification of the excitation mechanism, been conducted. Only in the case of sodium, because of its importance in spectral line-reversal studies, has sufficient work been carried out for the mechanism to be fully understood. In all others, the interpretation of the experimental data remains open to question.

Sodium

Considerable effort has been devoted to studying the excitation of sodium in shock waves, mainly because of the value of sodium line-reversal techniques for measuring shock temperatures. From his studies on flames, Gaydon[11] suggested that energy transfer to sodium atoms occurred predominantly from vibrationally-excited molecules so that the sodium excitation temperature provides a measure of the vibrational temperature of the molecules. On this basis, the slow rise of the sodium temperature behind shock waves in nitrogen was interpreted as being due to the vibrational relaxation of the nitrogen molecules[12]. The relaxation times measured in this way appeared to agree very closely with those obtained using other techniques[13] and hence provided strong evidence for the proposed mechanism. More recently, Millikan and White[14] have reexamined the high temperature relaxation of nitrogen by two independent methods and have demonstrated that the previous measurements, and therefore also those based on sodium line-reversal, were in error.

Hurle[15] has conducted a very careful reexamination of the sodium line-reversal method for measuring vibrational relaxation times in nitrogen. The results demonstrated conclusively that the sodium excitation temperature and the nitrogen vibrational temperature remain identical during the relaxation and that the discrepancies obtained previously could be attributed almost entirely to approximations made during the data analysis, notably that the vibrational temperature rather than the vibrational energy had been used as a monitor of the relaxation. Perhaps the most convincing demonstration of the correlation with the vibrational temperature was obtained from the early stages of the low-temperature experiments, in which the translational and rotational temperatures of the nitrogen would have been sufficiently high to produce a detectable amount of emission but, since the vibrational mode of the nitrogen had not relaxed, none was observed.

Estimates of the excitation and quenching cross-sections ($> 3 \cdot 5 \times 10^{-16}$ cm^2) in comparison with the momentum transfer cross-sections, show that the excitation probability per collision from vibrationally-excited nitrogen is close to unity. An efficiency of this magnitude for energy transfer suggests a near-resonant process and it appears likely that transfer occurs from nitrogen in the $v'' = 7$ or 8 levels.

$$\text{Na}(^2S) + \text{N}_2(v'' = 7, 8) \rightarrow \text{Na}(^2P) + \text{N}_2(v'' = 0)$$

The $\text{Na}(^2P)$ level lies 17,000 cm^{-1} above the ground state and the N$_2$ levels 15,700 cm^{-1} and 17,200 cm^{-1} above the $v'' = 0$ state so that the energy matching is quite close. Conclusive evidence for population of the excited state of the sodium atom by vibrationally-excited nitrogen molecules (Process II) thus appears to have been presented.

Studies of sodium excitation temperatures in expanded nozzle flows following shock wave heating have revealed that an alternative mechanism may apply in certain instances[16]. Because argon cannot readily excite sodium atoms, for the reasons given above, the excitation temperature behind the shock front should lie below the equilibrium value, but instead line-reversal measurements have indicated anomalously high values in this region. The electron temperatures were also shown to be very high and the upper levels of sodium are apparently populated by the free electrons present in the system. An estimate of the excitation cross-section for electrons gave a value about ten times that for vibrationally-excited nitrogen molecules. Therefore, in the absence of diatomic and polyatomic molecules, the dominant excitation mechanism appears to involve transfer of energy from free electrons (Process V). Because such a mechanism is unlikely to be specific to sodium atoms, and because the transfer from vibration is a near-resonant process of high efficiency, it appears reasonable to make the generalization that excitation by electrons will be more efficient than vibrational energy transfer for all atomic species.

When one per cent nitrogen was added to the argon it was found that the sodium excitation followed the nitrogen vibrational temperature. This is explained by the efficient thermalization of the electrons by the nitrogen molecules. One would therefore expect excitation by electrons to become of lower significance whenever diatomic and polyatomic species are present.

Chromium

Line reversal techniques are always open to criticism because the additive may be involved in a chemical interaction with the system under investigation. For this reason, confirmation of the sodium line-reversal results has been sought by using one of the chromium atomic spectral lines as a temperature monitor. The chromium temperatures have been found to agree very closely with those obtained using sodium[17]. However, a recent study[18] of the $Cr(^7P_2 \rightarrow {}^7S_3)$ transition using shock-tube techniques illustrated the difficulties incurred in attempting to elucidate mechanisms for thermal excitation of electronic states.

The excited chromium atoms were generated by shock-heating chromium carbonyl in a large excess of argon. Examination of the time-resolved intensity records of the 4289 Å line at temperatures above 2000°K revealed the participation of two distinct excitation processes. One of these produced a rapid increase in the emission intensity followed by a similar rapid decay, whilst the other gave an intensity record which displayed a slow rise and persisted for a longer period. The second process very probably corresponds to direct thermal excitation by argon atoms (Process I) which would be expected to yield a slow rise to an equilibrium situation.

The first process is more difficult to explain. One possibility is that the decomposition of chromium carbonyl leads to chromium atoms in both the ground and excited states. The rise in intensity would then correspond to the decomposition of the carbonyl and the fall to a decay in the excited state concentration by radiation and by collisional quenching. An alternative, although related, explanation is that the initial decomposition produces vibrationally-excited carbon monoxide molecules which then transfer energy to an electronic state of chromium. The latter explanation has some merit because of the known efficiency of vibrational–electronic energy transfer (Process II).

Direct population of excited states on dissociation has been invoked previously to explain anomalous emission behaviour in shock wave experiments. However, it seems unlikely that a reaction which is strongly endothermic [$\Delta H^0_{298} = 162$ kcal/mole for dissociation of $Cr(CO)_6$][19] could possibly lead to appreciable product formation in excited states and so alternative explanations of the experimental observations must be sought.

Inert Gases

A large number of investigations has been carried out on the emission characteristics of strong shock waves in the inert gases, helium, neon, argon, krypton, and xenon[1,2,3,4,20,21]. In all cases, the spectra consist of a large number of atomic lines superimposed on a continuum. Most of the studies have been concerned with the continuous emission and only in the case of xenon have detailed investigations been carried out on the kinetic behaviour of the atomic line transitions.

The majority of the work has been carried out at sufficiently high temperatures for appreciable ionization to occur ($>10,000°$K) and the continuum radiation has been ascribed to radiative ion–electron recombination (Process VI) rather than to bremsstrahlung (Process V). As the emission intensity provides a measure of the degree of ionization, the increase of intensity with time has been used to give rates of ionization. Ionization rates can be measured by other means and in most cases the different techniques provide similar results, thus confirming the mechanism leading to continuum emission.

Continuum emission is not in itself relevant to the present discussion, since the transition involved is normally from a repulsive potential energy curve, which does not therefore represent a stable electronic state, to the ground state of the atom. However, if ion–electron recombination is occurring then one would also expect transitions to stable excited electronic states, either by radiative or by three-body recombination.

$$A^+ + e \rightarrow A^* + h\nu$$

$$A^+ + e + M \rightarrow A^* + M$$

These stable states would then produce discrete spectra corresponding to transitions to lower electronic states. This almost inevitably is the explanation of the atomic line spectra which are superimposed on the continua.

An even more interesting effect has been observed during the ionization relaxation process. In essentially all cases of significance, the temperature dependence of the rate of ionization corresponds not to the ionization potential of the atom but to the energy of the first excited state. The most recent investigation[22] gives values of 11.2–11.8 eV for argon compared with an excited state energy of 11.5 eV, 9.6–10.1 eV for krypton compared with 9.9 eV, and 8.2–8.4 eV for xenon compared with 8.3 eV. The reason for this finding is that, at least in the initial stages, ionization occurs by a stepwise mechanism involving atom–atom collisions. Since the energy required to form the lowest excited state is much greater than that for subsequent transitions (in argon, 11.55 eV compared with 4.21 eV), the rate of population of the first excited state therefore determines the overall rate. For the mechanism to be feasible, these intermediate excited states must be relatively long-lived. This suggests that the metastable states rather than the resonance states, which would undergo rapid radiative decay to the ground state, are involved. However, as the gas is optically thick to the resonance radiation, trapping[23] will occur and so both types of electronic states may be involved. The energy differences between such states are too small to enable a distinction to be made on the basis of the existing experimental evidence.

The ionization rate measurements are usually expressed in terms of cross-section coefficients and yield values of 1.2×10^{-19} cm^2/eV, 1.4×10^{-19} cm^2/eV and 1.8×10^{-20} cm^2/eV for argon, krypton and xenon respectively. Since the temperatures involved correspond to energies of the order of an electron volt, these figures correspond roughly to 10^{-4} of the momentum transfer cross-sections. Thus atom–atom excitation is much less efficient than vibrational–electronic energy transfer, which gives values similar in magnitude to the gas kinetic cross-sections. These cross-sections are also found to be about one-fortieth of the electron–atom cross-sections. The reason for the order-of-magnitude difference between xenon and the other inert gases is not known. However, the cross-section for argon atoms ionizing xenon has been found to be identical to that for xenon ionizing xenon so that the difference does not lie in the projectile species but in that undergoing excitation[24].

It must be noted that this ionization mechanism applies only at low temperatures and at very low impurity levels. If these conditions are not satisfied, the electron concentrations build up sufficiently rapidly for direct electron–atom excitation to occur and the formation of intermediate excited states may not be required.

In the case of xenon, an extensive study has been made of the light emission itself[21] as well as of the ionization relaxation behaviour[25]. Unfortunately, the investigators disagree[26] as to the proper interpretation of their results and, in consequence, it seems worthwhile to summarize only the most salient features of their mechanisms. Atomic line spectra and continuous spectra are observed in both the visible and ultraviolet regions. The time dependence of the emission intensity in the two regions differs, in the visible decaying rapidly after the initial rise whilst in the ultraviolet persisting with little reduction. The authors appear to agree that the decay in appearance of the luminosity cannot be explained by ion–electron recombination or by bremsstrahlung (Processes V and VI), and prefer the suggestion that the metastable xenon atoms recombine by three-body processes to give a stable excited diatomic molecule. The continua are then due to radiative dissociation of two free xenon atoms

$$Xe^* + Xe^{(*)} + M \rightarrow Xe_2 + M$$

$$Xe_2 \rightarrow Xe + Xe + hv$$

Where the authors disagree most strongly is in the electronic states involved in these reactions and in the actual transitions involved in radiative cooling.

Mulliken[27] has carried out semi-empirical calculations which predict the existence of stable molecular states of xenon based on excited atomic levels. Since the same calculations also predict the existence of a stable molecular ion and this species has been observed experimentally[28], these predictions deserve very serious consideration. Given their existence, the population of such states would almost certainly take place under shock wave conditions. The reactions which would ensue are not appreciably different from those which occur in the bromine system, described below, except, of course, that the ground states of the atoms have only a repulsive interaction.

In summary, excited electronic states of inert gas species can be populated by direct thermal means (Process I), by ion–electron recombination (Process VI), and probably by association of other excited states, a mechanism which really constitutes a form of chemi-excitation (Process IV).

Bromine

Although bromine is a molecular species, it is considered here because, during thermal excitation, it dissociates into atoms and it is these atoms which are responsible for the emission of light. Another reason for its inclusion is that the mechanism compares very closely with that suggested above for emission in xenon.

Dissociation of ground-state ($^1\Sigma_g^+$) bromine molecules is expected to produce bromine atoms in the $^2P_{3/2}$ state although these may readily be excited to the $^2P_{1/2}$ state 10 kcal/mole higher. Three excited molecular species may be formed by association of these atoms. These can then decay to the ground state by radiative transitions so that the overall process is a chemiluminescence produced by two-body association (Type I):

$$Br(^2P_{3/2}) + Br(^2P_{1/2}) \rightarrow Br_2(^3\pi_{0^+u}) \rightarrow Br_2(^1\Sigma_g^+) + hv$$

$$Br(^2P_{3/2}) + Br(^2P_{3/2}) \rightarrow Br_2(^3\pi_{1u}) \rightarrow Br_2(^1\Sigma_g^+) + hv$$

$$Br(^2P_{3/2}) + Br(^2P_{3/2}) \rightarrow Br_2(^1\pi_u) \rightarrow Br_2(^1\Sigma_g^+) + hv$$

Palmer[29] investigated the continuum emission from shock-heated bromine by measuring the intensity as a function of both temperature and atom concentration at three different wavelengths. The temperature coefficients were predicted for each wavelength from the potential energy curves (Figure 2.3) by applying the Franck–Condon principle to the

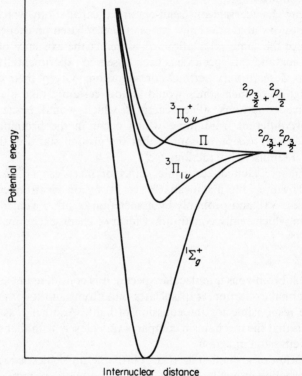

Figure 2.3   Potential energy curves for bromine (schematic only)

appropriate vibrational wave functions. In this way, Palmer was able to demonstrate that all three processes contributed to the emission and, therefore, that electronically-excited levels of bromine were populated predominantly by this dissociation-recombination mechanism. At temperatures where appreciable dissociation occurs, the formation of excited molecular states in this manner can normally be expected. The extent to which prior population of excited atomic levels takes place, as in the present instance, is more difficult to predict.

2.3 EXCITATION OF MOLECULAR SPECIES

It was noteworthy in the discussion above that chemical excitation processes are frequently involved in electronic excitation of atomic states. In the case of molecules, excitation occurs more commonly by chemical routes and other excitation mechanisms are of lesser importance. The examples below have therefore been selected to illustrate the various types of chemical reaction which can lead to excitation.

C_2, CN, and CH

In addition to the persistent atomic line spectra, emission from the C_2 and CN radicals is a feature of many shock wave investigations[30] despite the precautions which are normally taken to minimize contamination by the hydrocarbon impurities which are responsible for the production of these species.

C_2 appears on shock-heating hydrocarbons, carbon tetrachloride, carbon monoxide, and cyanogen[31]. The comparable ease with which C_2 emission occurs from both methane and acetylene indicates that the mechanism involves association of radicals containing a single carbon atom rather than stripping of hydrogen atoms from a C_2 hydrocarbon[32]. Molecules containing two carbon atoms are known to be produced during the oxidation of methane[33]. Experiments with isotopically-labelled acetylene[34] have shown that the two carbon atoms in C_2 arise from different molecules and thus confirm that the carbon–carbon bond ruptures during the formation of excited C_2. Further confirmation of this argument is provided by the finding that the initial excitation temperature of C_2, obtained by simultaneous emission and absorption measurements, is initially high and then falls to the thermodynamic equilibrium value[32]. The initial emission therefore appears to be chemiluminescent in origin and presumably follows an association or exchange process of the type listed above. It is very unlikely that association of free carbon atoms is involved since such atoms are not detected in high temperature systems of this type. A stronger possibility is that exchange of hydrocarbon radicals is involved, typified by the reaction

$$CH + CH \rightarrow C_2{}^* + H_2$$

The CH radical is frequently observed in luminous flames but less commonly in the shock tube. Since all these excited species apparently arise from chemical processes, this disparity may well be due to the different chemistry in the two systems, and in particular to the high concentration of oxygen-containing species which are present in flames. Thus, the presence of excited CH during the oxidation of acetylene can be attributed, on kinetic arguments[35], to the reaction

$$C_2H + O \rightarrow CO + CH^*$$

and CH* is believed to be formed in flames[36] by the reaction

$$C_2 + OH \rightarrow CH^* + CO$$

The emission from C_2 in shock-heated hydrocarbons is certainly reduced by the addition of hydrogen and oxygen together, although the two gases do not have the same effect when added separately, which adds weight to the argument that the OH radical is responsible. However, as a simultaneous appearance of CH emission is not observed, the evidence may not be regarded as conclusive.

The pyrolyses of cyanogen and of BrCN[37] both yield C_2 and CN emission although, in these examples, the CN appears before the C_2. This indicates that CN forms by direct dissociation and that the C_2 appears as a result of an analogous exchange reaction to that proposed above:

$$CN + CN \rightarrow C_2 + N_2$$

On this basis, the CN should be excited by a thermal mechanism rather than by chemi-excitation. Supporting evidence for this is provided by the observation[32] that the excitation temperature of the CN commences at a low value close to the front and then rises to the equilibrium value. It is interesting to find an excited molecular level which is apparently populated by direct thermal means. In the same experiments, the C_2 has a high initial excitation temperature indicating that the reaction above leaves the C_2 with a non-equilibrium energy distribution. A surprising outcome of this investigation is that the addition of a trace of acetylene to the cyanogen produces a high excitation temperature for both C_2 and CN. It is difficult to give a detailed interpretation of this result but it certainly indicates that the excited state of CN is not populated exclusively by thermal means.

Even for C_2, the excitation mechanism is by no means fully understood. The radical exchange process appears reasonable in the examples quoted above but it is difficult to suggest a comparable route which will explain the occurrence of excited C_2 on shock-heating carbon monoxide[31]. Also, in contrast to the mechanisms proposed in the preceding discussion, a recent shock tube investigation of the pyrolysis of acetone[38] has provided evidence for the direct thermal excitation of C_2. This deduction is based on the observed first-order pressure dependence and on the absolute intensity of the emission so that the arguments appear convincing. However, until more information is gained on the mechanism by which the ground state species is formed, it may be dangerous to regard the evidence as conclusive, at least with respect to other systems. Certainly the experiments involving isotopically-labelled molecules and, indeed, consideration of likely processes involved in the pyrolysis of acetone, clearly show that bimolecular reactions are involved in C_2 formation.

As mentioned above, CH is an infrequent emitter in shock wave experiments. However, it does appear in certain cases, for example on shock-heating methyl bromide[37] (together with C_2 emission). This confirms that the detailed chemistry of the system must be responsible for determining whether CH emission is significant.

Nitrogen

The emission from shock-heated nitrogen, either alone or in the presence of oxygen, has been extensively studied[39]. The time-dependence of the emission is interesting: the intensity rises rapidly behind the front, 'overshoots' the equilibrium value and then decays slowly to it. In this example, it appears that the excited state population must be strongly coupled to the 'translational' temperature of the gas so that an equilibrium between the ground and excited electronic states is achieved before vibrational and/or chemical equilibration have occurred.

This qualitative explanation has been examined in greater detail by Hammerling, Teare and Kivel[40] and the treatment is interesting in that it helps to show which of the various processes are critical in determining the magnitude of the overshoot. The chemistry of the system was assumed to depend on two processes: the dissociation and recombination of nitrogen

$$N_2 + M \rightleftharpoons 2N + M$$

and the vibrational relaxation of the nitrogen molecules. The rate constant for nitrogen dissociation at high temperature was estimated from low-temperature results, assuming that the behaviour with temperature paralleled that observed with oxygen. It was assumed that nitrogen atoms possess a third-body efficiency five times that of nitrogen molecules, also by analogy with oxygen.

The use of a dissociation rate constant in this manner may not be valid because vibrational equilibrium may not have been achieved during the dissociation. Account has been taken of this by allowing dissociation to occur from any vibrational level, at a rate dependent on the energy change involved, and then coupling the dissociation to the vibrational relaxation.

The actual emission was due to the First Negative, $N_2^+(B^2\Sigma_u^+) \rightarrow N_2^+(X^2\Sigma_g^+)$, system and it was assumed that N_2^+ ions were formed by associative ionization

$$N + N \rightarrow N_2^+ + e$$

and by charge exchange

$$N^+ + N_2 \rightleftharpoons N_2^+ + N$$

The population of the excited state of N_2^+ was considered either to be in local equilibrium with the translational temperature or to occur via transfer from vibrationally-excited species during the relaxation. A typical solution, which shows the effect of the various alternatives, is shown in Figure 2.4. The experimental data so far obtained appear to correlate most closely with the curves based on coupling of vibration and dissociation, with excitation taking place via vibration–electronic energy transfer.

Although the computer simulation of the emission characteristics is very instructive and helps to demonstrate the types of processes which may lead to 'overshoot' phenomena, it is evident that more information, particularly of a quantitative nature, is required concerning the individual steps involved.

In shock-heated nitrogen, emission has been found to originate predominantly from two systems: the First Positive,

$$N_2(B^3\pi_g) \rightarrow N_2(A^3\pi_u^+)$$

and the First Negative,

$$N_2^+(B^2\Sigma_u^+) \rightarrow N_2^+(X^2\Sigma_g^+)$$

Since the appearance of the former involves a change in multiplicity and of the latter, the formation of molecular ions, it is likely that the mechanisms for both involve an initial dissociation into nitrogen atoms. For this reason, an investigation[41] was carried out on the effect of shock-heating nitrogen containing a high concentration of nitrogen atoms produced by a pulsed electric discharge. The intensity of the Lewis–Rayleigh afterglow

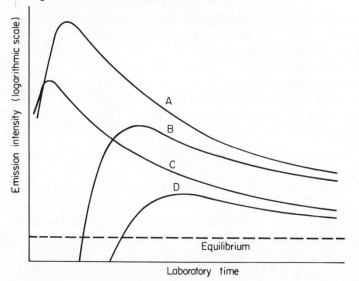

Figure 2.4 Predicted emission intensity profiles from shock waves in nitrogen (schematic only). A. Coupled vibration and dissociation, equilibrium excitation. B. Coupled vibration and dissociation, excitation by vibrational–electronic energy transfer. C. Parallel vibration and dissociation, equilibrium excitation. D. Parallel vibration and dissociation, excitation by vibrational–electronic energy transfer

resulting from atomic recombination was used as a measure of the nitrogen atom concentration.

On the basis of the results of this work, the following mechanism has been proposed for the excitation of the $N_2(1+)$ system:

$$N_2(X) + N \rightarrow N_2(A) + N \qquad (1)$$

$$N_2(X) + N_2 \rightarrow N_2(A) + N_2 \qquad (2)$$

$$N_2(A) \rightleftharpoons N_2^\dagger(B) \qquad (3)$$

(3) is fast and is therefore assumed to be in local equilibrium at the translational temperature of the system. Process (1) is found to be a hundred times as efficient as process (2) in promoting excitation of the $N_2(A)$ state. The experimental evidence cannot completely eliminate direct excitation of the $N_2(B)$ state, but three-body recombination into the (B) state can be ruled out because it would predict a different temperature

dependence. The rate constant expression obtained for process (1) was
$k_N = 1.9 \times 10^{-6} T^{-3/2} \exp(-E_{X \to A}/kT) \, \text{cm}^3/\text{molecule s}$ which gives an
effective cross-section of $2.7 \times 10^{-18} \, \text{cm}^2$ at $12{,}000°\text{K}$. Since the
$N_2(X) \to N_2(A)$ transition involves a change in multiplicity, the greater
efficiency of nitrogen atoms compared with nitrogen molecules in promoting the transition may be because they are better able to facilitate the
change of spin.

For the excitation of the $N_2^+(1-)$ system, the sequence

$$N + N \to N_2^+(X) + e$$
$$N_2^+(X) \rightleftharpoons N_2^+(A) \rightleftharpoons N_2^+(B)$$
$$N_2^+ + N \rightleftharpoons N_2 + N^+$$

is suggested. As before, the excited states are assumed to be in local
equilibrium at the translational temperature. The mathematical treatment is based on the assumption that associative ionization involves
ground state (^4S) atoms.

It is unlikely that either the singlet or the triplet potential energy curves
from the ground state intersect that of the $N_2^+(X)$ state and cross-over
from the $^7\Sigma_u^+$ state would not conserve electron spin, so that one of the
atoms involved in the associative ionization is probably itself in an
excited state. Gilmore[42] has discussed this matter in some detail and
suggests that the nitrogen atom will be in the ^2D or ^2P state. Unfortunately the rate of excitation of nitrogen atoms into these states is unknown so that it is not possible to determine whether a sequence involving
excited atoms would be sufficiently rapid to explain the observed rates of
excitation of the higher molecular states.

Both these mechanisms are of interest because they are not strictly
covered by the classification in the introduction. In the excitation of the
$N_2(B)$ state, the process directly responsible appears to be thermal in
nature although the exciting species is not a stable molecule but is an atom
formed by a prior chemical reaction. Thus the process cannot be defined
as chemi-excitation despite the fact that the greater efficiency of energy
transfer when atoms rather than molecules are involved must be associated with chemical 'forces' in the interaction potential. It has already
been suggested that as molecular complexity increases so does the
probability of chemi-excitation versus thermal excitation increase. This
example illustrates that the reason may often lie simply in the stronger
coupling between molecular and electronic motions which occurs when
chemically-reactive intermediates are generated.

The associative ionization process involved in molecular ion formation is not so different from mechanisms discussed previously and, in fact, bears close comparison with the sequence suggested to explain emission from shock-heated bromine.

At temperatures of about 9000°K in nitrogen, the molecules will be almost entirely dissociated into atoms but will only be partially ionized. The most prominent molecular emission arises from the First Negative system of N_2^+ but the atomic ions will also be responsible for emission either by deceleration (Process V) or capture (Process VI) of electrons. The intensity of radiation from these two sources can be calculated theoretically[43] and it is found that there is a discrepancy between the total intensity predicted and that observed experimentally[44]. This has been attributed to interactions between the neutral atoms and the electrons. The free–free scattering is insufficient to account for the whole of the discrepancy and the additional contribution is probably due to the formation of N^- ions. Comparison with studies using electric arcs[45] suggests that the atoms involved are initially in the excited ^2D state and lead to negative ions in the ^1D state. The ground-state has a very low binding energy and is therefore not expected in significant amounts. Although the density dependence is in agreement with this mechanism, a more complete interpretation both of this and the other sources of emission really requires a more detailed analysis of the variation of intensity during the approach to equilibrium.

Sulphur Dioxide

A very detailed investigation of the emission from shock-heated sulphur dioxide has been conducted by Levitt and Sheen[46]. The emission is found to lie in the range 2500–5000 Å and to involve contributions from the $^3B_1$ state, the $^1B_2$ state, and a third higher electronic state, possibly the $^1B_1$. Since the emission intensity reaches a maximum close to the shock front, at a stage when only a negligible amount of dissociation can have occurred, the excited states must be formed directly from ground state SO_2 molecules rather than by recombination of SO and O. On this basis, the excitation of sulphur dioxide parallels more closely the behaviour observed for atomic species rather than that experienced by molecular species. The results can, in fact, all be explained in terms of simple collisional excitation and quenching

$$SO_2 + M = SO_2^* + M$$
$$SO_2^* = SO_2 + h\nu$$

In the case of the triplet ($^3B_1$) species, cross-over from vibrationally excited ground-state molecules or from excited singlet ($^1B_2$) molecules is also possible but unlikely assumptions about the rates of the various processes would be required to obtain the observed kinetics. Use of nitrogen rather than argon as diluent demonstrates that the intensity of emission, and therefore presumably the rate of excitation, depends on the translational temperature of the diluent gas.

The temperature coefficient of the emission intensity at 4360 Å is found to be 73·2 kcal, which is close to the energy of the $3B_1$ state of 73·5 kcal. This demonstrates that the emission arises from the lowest vibrational level of the upper state. The temperature coefficient at the long wavelength cut-off (4900 Å) is a little above 73·2 kcal/mole suggesting that emission in this region occurs from vibrational levels above the lowest. This implies that the *initial* excitation may not necessarily be into the ground vibrational level. An interpretation of this observation would require a detailed knowledge of the potential energy surfaces of the electronic states involved. The increased temperature coefficient measured at wavelengths below 4360 Å is attributed to participation of emission from the singlet ($^1B_2$) state, which is responsible for the bulk of the emission at 3800 Å. At shorter wavelengths, the emission shows a similar temperature coefficient to that determined for the $^3B_1$ state and is believed to arise from a third electronic state.

In summary, it appears that, with this particular molecule, either Process I or Process II is responsible for the population of the excited electronic state. It is doubtful whether there is any value to be gained in trying to distinguish between the two mechanisms, since if a vibrationally-excited intermediate is involved the crossover must be very efficient and one is not concerned here with vibrational–electronic energy transfer between dissimilar species as in the case of sodium.

Although the evidence points to direct thermal excitation, the situation may differ in more complex systems. In the oxidation of hydrogen sulphide, for example, where an induction delay is observed before SO_2 is produced, the onset of emission from excited SO_2 lags significantly behind the appearance of ground-state SO_2 in absorption[47]. Since the atom and radical concentrations are high in branching-chain reactions of this type, the emission may readily be attributed to radiative recombination of SO and O as in the flash photolysis studies[48]. The recombination emission can even be detected in shock-heated sulphur dioxide at wavelengths below 2600 Å, the mechanism being clearly differentiated by the slow rise in emission intensity following the passage of the shock front. Even with sulphur dioxide, therefore, the mechanism of electronic excitation is intimately concerned with the chemistry of the system.

Nitrogen Dioxide

Levitt[49,50] has undertaken a detailed study of the emission from shock-heated nitrogen dioxide on very similar lines to that carried out on sulphur dioxide. Once again the emission appears to be thermal in origin and the intensity is related to the translational temperature of the gas. Nitrogen and argon display identical efficiencies in exciting the spectrum which indicates that, as in the case of SO_2, the rotational energy of nitrogen is probably not available for exciting NO_2 and demonstrates further that the excitation is not coupled to the vibrational temperature of the nitrogen.

The variation of intensity with composition indicates that the excited state, denoted by NO_2^*, is quenched by ground-state NO_2 by a mechanism for which the reverse is ineffective. This implies that one of the two reactions

$$NO_2^* + NO_2 \rightarrow NO_3 + NO$$
$$NO_2^* + NO_2 \rightarrow 2NO + O_2$$

must be responsible.

The variation of emission intensity with wavelength can be interpreted in terms of a Boltzmann distribution for the population of the upper vibrational levels. Since dissociation may be expected to occur from the highest levels and the rate of vibrational energy transfer could prove inadequate to counteract the resulting depopulation, some deviation from a strict Boltzmann distribution might be anticipated. Support for this contention is provided by the observation that, at the higher temperatures, an appreciable depopulation occurs of levels close to the pre-dissociation limit.

In the case of sulphur dioxide, it was noted that under appropriate conditions, for example flash photolysis or shock wave heating of hydrogen sulphide–oxygen mixtures, the emission could arise from a recombination mechanism rather than from direct thermal excitation. Studies on flames with added nitric oxide[51] and on shock-heated air[52] have demonstrated that a corresponding alternative mechanism may also lead to the electronic excitation of nitrogen dioxide:

$$O + NO + M \rightarrow NO_2' + M$$
$$NO_2' \rightarrow NO_2 + h\nu$$

The most recent study[50] has revealed that the spectral characteristics of the two sources differ. Although both spectra appear continuous, the maximum intensity of the thermal emission appears to lie in the infrared region beyond 9500 Å while the peak intensity of the recombination spectrum lies in the visible at about 6000 Å. In addition, the radiative lifetime of the thermal emission is found to be about 10^{-7} s which is close to that calculated for the visible absorption spectrum $(2 \times 10^{-7}$ s) but is quite different from the value obtained for the fluorescence spectrum $(4.4 \times 10^{-5}$ s).

Because of this evidence, the excited state formed by the recombination mechanism has been denoted by NO_2'. The question is naturally raised as to whether NO_2^* and NO_2' refer to different electronic states of nitrogen dioxide. A possible explanation for the conflicting data could be that the different mechanisms lead to very different vibrational level distributions within the same state. Despite the depopulation effect referred to above, there is evidence that vibrational relaxation would be essentially complete within the collisional lifetime of the radiating species. It therefore seems likely that two different electronic states are involved. This system may differ from that of sulphur dioxide where, at present, the evidence is conflicting as to whether identical electronic states are involved in both excitation mechanisms.

The recombination spectrum[53] has been studied independently of the thermal emission. Use of this data in combination with the shock tube results enables the total emission behaviour to be calculated for systems which permit both mechanisms to operate, e.g. shock-heated air, flames with added NO. For the spectral region between 4000 and 12,000 Å, the majority of the emission from shock-heated air appears to arise from the recombination of O and NO.

It will be evident from the examples discussed above and those which follow that thermal excitation (Process I) and chemi-excitation (Process IV) are normally the only two mechanisms which need to be considered in discussing the population of molecular electronic states. The example of shock-heated air provides direct evidence for a suggestion made elsewhere, namely that both mechanisms may operate simultaneously and with comparable importance in the same system. It serves therefore to emphasize the difficulties inherent in attempting to interpret emission measurements on shock-heated gases.

OH

The combustion of hydrogen in oxygen leads to the emission of ultra-violet light which can be identified with the $^2\Sigma^+ \to {}^2\pi$ transition of the hydroxyl radical[54]. The characteristics of the emission[55], and in particular the high energy involved, appear to rule out direct thermal population of the upper state and indicate instead that a chemiluminescent reaction must be sought to explain the emission.

The study of this emission has formed the subject of two shock tube investigations[56]. Six possible reactions could lead to the chemiluminescent emission

$$H + H + OH \to OH^* + H_2$$

$$O + O + OH \to OH^* + O_2$$

$$H + OH + OH \to OH^* + H_2O$$

$$O + H_2 + OH \to OH^* + H_2O$$

$$O + H + M \to OH^* + M$$

$$H + O_2 + H_2 \to OH^* + H_2O$$

If the course of the hydrogen–oxygen reaction is followed, for example by monitoring the absorption due to OH, several different regions may be distinguished. Due to the branching-chain character of the reaction, an induction period is observed followed by an exponential growth in radical concentration. The rapid bimolecular reactions eventually reach a partial-equilibrium, at which the radical concentrations attain a steady value. These high radical concentrations then decay due to the slower third-order recombination processes becoming important.

The rate constants for all the major reactions are now known with some precision and it is possible to compute the partial equilibrium composition from the shock strength and the initial composition. Comparisons of the predicted intensity dependences on partial equilibrium composition and temperature with the experimental measurements completely eliminate the first four reactions as possible sources of the emission.

The earlier investigation provided kinetic evidence for

$$H + O_2 + H_2 \to OH^* + H_2O$$

being the important reaction, but comparison of the exponential growth rate for the emission with that observed for the hydroxyl radical concentration demonstrates that the process leading to emission is second order with respect to the concentration of chain carriers. By a process of elimination, therefore, the excited OH radical is formed by the reaction

$$H + O + M \rightarrow OH^* + M$$

Recombination of atom or radical species to give an excited state is perhaps the commonest source of chemiluminescence and several examples have already been quoted. What is perhaps more interesting is the complex chemistry involved in a system which is probably the simplest example of a branching-chain oxidation. In consequence, a very detailed analysis becomes necessary in order to determine the actual excitation mechanism.

In general, high excitation energies (in the ultraviolet) will appear only when the total number of chemical bonds increases and the number of moles of reactant falls. Furthermore, this behaviour is indicative of high atom and radical concentrations, such as are found in flames and detonations, and therefore is a consequence of branching-chain reactions of this type. Very few systems apart from those involving chain-branching reactions are accompanied by emission of light with both high energy and high intensity.

NH

Emission characteristic of the NH radical is frequently observed from high temperature gases containing nitrogen and hydrogen[36,57,58] due in part to the high oscillator strength of the $^3\Pi \rightarrow {}^3\Sigma^-$ transition[59]. The radiation has been studied by shock wave techniques, using a number of different nitrogen and hydrogen sources, and by a variety of other techniques. So far, however, there is only very limited agreement as to the mechanism of excitation of the $^3\Pi$ state.

On examination of the spectra from reflected shocks, Avery, Bradley and Tuffnell[60] were led to suggest that the excitation of the NH radical could not be thermal. This conclusion was based on the observation that the temperature coefficient of the emission maximum differed from the predicted value, on the absence of vibrational equilibration in the excited state, and on the kinetic characteristics of the decay of the emission. Cann and Kash[61] subsequently carried out a similar investigation under almost identical conditions and their findings disagreed considerably with those of the previous workers. It is difficult to ascertain which of the various results should be taken as correct. Certainly both groups of workers were confronted by a very complex array of experimental data and were necessarily forced to make somewhat subjective assessments in isolating features which were common to all experiments. The early workers clearly oversimplified the decay characteristics and a reappraisal of their data, together with the results of subsequent measurements on different systems[62], agree at least qualitatively with Cann and Kash's suggestion of two simultaneous decay processes.

In spite of the numerous discrepancies, certain important features are common to both systems. The temperature dependence observed for the intensity maxima was not equal to the value predicted for thermal emission and none of the studies, apart from that involving hydrazoic acid[58], showed any population of the $v = 3$ state. It appears, therefore, that the discussion by Avery and coworkers of the possible mechanisms for populating the upper electronic state may still be meaningful.

The substance of these arguments is that it is difficult to find a simple reaction which is sufficiently exothermic to produce NH in a highly excited state but that association reactions leading to molecular nitrogen will be highly exothermic because of the extremely high bond energy of N_2. The suggestion which is made is that an excited state of nitrogen is produced which then transfers energy to the NH radical, very probably via a near-resonant process[63]. Association of NH radicals

$$NH + NH \rightarrow N_2* + H_2$$

must be considered as a strong possibility although alternative reactions cannot be ruled out. The excited state of nitrogen may well be a vibrationally excited $N_2(A^3\Sigma_u^+)$ molecule although once again other states of N_2 cannot be eliminated.

Strong evidence for this mode of excitation has been obtained from studies on the reaction of atomic nitrogen with atomic hydrogen[64] and on the photodecomposition of hydrazoic acid[65]. In the former, the excited nitrogen is expected to arise from atomic association reactions rather than from radical–radical reactions as in the shock wave studies, but otherwise the features of the observed emission and the reactions postulated are closely comparable.

It is noteworthy that the OH emission observed during the oxidation of hydrogen occurs by simple atom recombination[56]. The reason that the analogous reaction

$$N + H + M \rightarrow NH* + M$$

is less likely to be significant in the present instance is that, in the absence of branching-chain kinetics, the atom concentrations will be too low to permit a reaction which is second order in atom concentration to achieve any importance. Any similarity ends when the oxidation of ammonia, either by oxygen or nitric oxide, is considered. The time at which the NH emission appears is independent of the concentration of oxidant and of the induction time for the branching-chain reaction, as measured by the appearance of ground-state OH radicals[62]. The NH excitation depends entirely on the independent pyrolytic reactions of the N/H compound and is therefore not a characteristic of a branching-chain reaction.

2.4 EXCITATION OF SOLID PARTICLES

An extensive investigation of the emission spectra produced by shock-heating solid particles has been conducted by Nicholls and Parkinson[66,67,68]. The mechanism of excitation in such systems must also include the vaporization of the solid material.

Excitation of solid aromatic hydrocarbons[68] (e.g. benzene, naphthalene, anthracene) over a wide temperature range (3000–13,000°K) led to the same emission spectra, characteristic of C_2, CN and CH, as observed by shock-heating the gas-phase species directly. The mechanism therefore appears to include the two processes of evaporation and excitation consecutively.

In the case of inorganic solids, in particular metal oxides, where the heats of vaporization and the vaporization temperatures are much greater, a rather different type of behaviour was observed. At low temperatures ($\sim 2000°K$), only the characteristic molecular spectra were obtained; at intermediate temperatures ($\sim 5000°K$) both atomic and molecular spectra were excited, and at high temperatures ($\sim 10,000°K$) the spectra were entirely atomic[67]. From this and other types of experiments it has been concluded that the initial evaporation leads directly to atoms which then recombine at the lower temperatures to produce the excited diatomic molecules.

The molecular spectra obtained in this way include those of AlO, BaO, CrO, FeO, MgO, PbO, TiO, and ZnO. This method of exciting molecular spectra is important because it eliminates impurities from electrode materials which tend to contaminate so many other spectroscopic sources.

REFERENCES

1. E. A. McLean, C. E. Faneuff, A. C. Kolb and H. R. Griem, *Phys. Fluids*, **3**, 843 (1960).
2. E. B. Turner, *Proc. N.S.F. Conf. on Stellar Atmospheres*, Indiana University, 1954.
3. F. H. Mies, *J. Chem. Phys.*, **37**, 1101 (1962).
4. H. E. Petshek, P. H. Rose, H. S. Glick, A. Kane and A. Kantrowitz, *J. Appl. Phys.*, **26**, 83 (1955).
5. V. N. Alyamorskii and V. F. Kitaeva, *Opt. Spectry.* (*USSR*) (*English Transl.*), **8**, 80 (1960).
6. R. N. Hollyer, A. C. Hunting, O. Laporte and E. B. Turner, *Nature*, **171**, 395 (1953).
7. E. B. Turner, *Dissertation*, Univ. of Michigan (1956).
8. A. F. Haught, *Phys. Fluids*, **5**, 1337 (1962).
9. O. Laporte and T. D. Wilkerson, *J. Opt. Soc. Am.*, **50**, 1293 (1960).
10. J. C. Clouston and A. G. Gaydon, 'The shock tube as a source for studies of emission and absorption spectra', *Proc. Conf. on Molecular Spectroscopy*, London (1958).
11. A. G. Gaydon, *Nat. Bur. Std. Circ. No. 523*, 1 (1954).
12. J. G. Clouston, A. G. Gaydon and I. I. Glass, *Proc. Roy. Soc.* (*London*), **A248**, 429 (1958); J. G. Clouston, A. G. Gaydon and I. R. Hurle, *Proc. Roy. Soc.* (*London*), **A252**, 143 (1959); A. G. Gaydon and I. R. Hurle, *Eighth Symposium* (*International*) *on Combustion*, Williams and Wilkins, Baltimore, 1962, p. 309; S. Tsuchiya and K. Kuratani, *Combust. Flame*, **8**, 299 (1964).
13. V. Blackman, *J. Fluid Mech.*, **1**, 61 (1956); S. J. Lukasik and J. E. Young, *J. Chem. Phys.*, **27**, 1149 (1957).
14. R. C. Millikan and D. R. White, *J. Chem. Phys.*, **39**, 98 (1963).
15. I. R. Hurle, *J. Chem. Phys.*, **41**, 3911 (1964).
16. I. R. Hurle and A. L. Russo, *J. Chem. Phys.*, **43**, 4434 (1965).
17. A. G. Gaydon and I. R. Hurle, *Proc. Roy. Soc.* (*London*), **A262**, 38 (1961); G. M. Kimber and D. H. Napier, *Combust. Flame*, **9**, 103 (1965).
18. W. S. Watts and S. H. Bauer, *J. Chem. Phys.*, **44**, 2206 (1966).
19. F. A. Cotton, A. K. Fischer and G. Wilkinson, *J. Am. Chem. Soc.*, **78**, 5168 (1956); F. A. Cotton, A. K. Fischer and G. Wilkinson, *J. Am. Chem. Soc.*, **79**, 2044 (1957); F. A. Cotton, A. K. Fischer and G. Wilkinson, *J. Am. Chem. Soc.*, **81**, 800 (1959).
20. G. E. Seay, L. B. Seely, Jr., and R. G. Fowler, *J. Appl. Phys.*, **32**, 2439 (1961).
21. W. Roth and P. Gloersen, *J. Chem. Phys.*, **29**, 820 (1958); W. Roth, *J. Chem. Phys.*, **31**, 844 (1959); P. Gloersen, *Phys. Fluids*, **3**, 857 (1960).
22. A. J. Kelly, *J. Chem. Phys.*, **45**, 1723 (1966).
23. T. Holstein, *Phys. Rev.*, **72**, 1212 (1947).
24. A. J. Kelly, *J. Chem. Phys.*, **45**, 1733 (1966).
25. H. S. Johnston and W. M. Kornegay, *Trans. Faraday Soc.*, **57**, 1563 (1961); W. M. Kornegay and H. S. Johnston, *J. Chem. Phys.*, **38**, 2242 (1963).
26. J. N. Bradley, *J. Chem. Phys.*, **32**, 1875 (1960); W. Roth, *J. Chem. Phys.*, **32**, 1876 (1960); W. Roth, *Phys. Fluids*, **4**, 788 (1961); P. Gloersen, *Phys. Fluids*, **4**, 789 (1961).

27. R. S. Mulliken, quoted by P. Gloersen, *Phys. Fluids*, **3**, 857 (1960).

28. J. A. Hornbeck, *Phys. Rev.*, **84**, 615 (1951); J. A. Hornbeck and J. P. Molnar, *Phys. Rev.*, **84**, 621 (1951); M. A. Biondi and L. M. Chanin, *Phys. Rev.*, **94**, 910 (1954).

29. H. B. Palmer and D. F. Hornig, *J. Chem. Phys.*, **26**, 98 (1957); H. B. Palmer, *J. Chem. Phys.*, **23**, 2449 (1955); H. B. Palmer, *J. Chem. Phys.*, **26**, 648 (1957).

30. E. B. Turner, *Phys. Rev.*, **99**, 633 (1955); R. J. Rosa, *Phys. Rev.*, **99**, 633 (1955); G. Charatis, L. R. Doherty and T. D. Wilkerson, *J. Chem. Phys.*, **27**, 1415 (1957).

31. A. R. Fairbairn and A. G. Gaydon, *Proc. Roy. Soc. (London)*, **A239**, 464 (1957).

32. A. R. Fairbairn, *Proc. Roy. Soc. (London)*, **A267**, 88 (1962).

33. J. E. Dove and D. McL. Moulton, *Proc. Roy. Soc. (London)*, **A283**, 216 (1965).

34. A. R. Fairbairn, *Eighth Symposium (International) on Combustion*, Williams and Wilkins, Baltimore, 1962, p. 304.

35. G. P. Glass, G. B. Kistiakowsky, J. V. Michael and H. Niki, *J. Chem. Phys.*, **42**, 608 (1965).

36. A. G. Gaydon, *The Spectroscopy of Flames*, Chapman and Hall, London, 1957.

37. E. F. Greene, *J. Am. Chem. Soc.*, **76**, 2127 (1954).

38. B. P. Levitt and N. Wright, *Trans. Faraday Soc.*, **63**, 282 (1967).

39. M. Camac, J. C. Camm, S. Feldman, J. C. Keck and C. Petty, *Inst. Aeron. Sci. Preprint*, No. 802, 26th Annual Meeting (1958); R. A. Allen, J. C. Keck and J. C. Camm, *Phys. Fluids*, **5**, 284 (1962); R. A. Allen, P. H. Rose and J. C. Camm, *Inst. Aeron. Sci.*, Paper No. 63-77, 31st Annual Meeting, New York (1963); R. A. Allen, *J. Quant. Spectry. Radiative Transfer*, **5**, 511 (1965).

40. P. Hammerling, J. D. Teare and B. Kivel, *Phys. Fluids*, **2**, 422 (1959).

41. K. L. Wray, *J. Chem. Phys.*, **44**, 623 (1966).

42. F. R. Gilmore, *Rand Corporation Memorandum*, RM-4034-PR (1964).

43. L. M. Biberman and G. E. Norman, *Opt. i Spektroskopiya*, **8**, 433 (1960).

44. R. A. Allen and A. Textoris, *J. Chem. Phys.*, **40**, 3445 (1964).

45. G. Boldt, *Z. Physik*, **154**, 330 (1959).

46. B. P. Levitt and D. B. Sheen, *J. Chem. Phys.*, **41**, 584 (1964); B. P. Levitt and D. B. Sheen, *Trans. Faraday Soc.*, **61**, 2404 (1965); B. P. Levitt and D. B. Sheen, *Trans. Faraday Soc.*, **63**, 540 (1967).

47. J. N. Bradley and D. C. Dobson, *J. Chem. Phys.*, **47**, 1555 (1967).

48. R. G. W. Norrish and A. P. Zeelenberg, *Proc. Roy. Soc. (London)*, **A240**, 293 (1957).

49. B. P. Levitt, *Trans. Faraday Soc.*, **58**, 1789 (1962); B. P. Levitt, *Trans. Faraday Soc.*, **59**, 59 (1963).

50. B. P. Levitt, *J. Chem. Phys.*, **42**, 1038 (1965).

51. W. E. Kaskan, *Combust. Flame*, **2**, 286 (1958).

52. W. H. Wurster and P. V. Marrone, *Study of Infrared Emission in Heated Air*, Cornell Aeronautical Laboratory Report QM-1373-A-4 (1961); W. H. Wurster and H. M. Thompson, *Bull. Am. Phys. Soc. II*, **10**, 278 (1965).

53. A. Fontijn, C. B. Meyer and H. I. Schiff, *J. Chem. Phys.*, **40**, 64 (1964).

54. K. J. Laidler, *The Chemical Kinetics of Excited States*, Oxford, 1955.

55. W. E. Kaskan, *J. Chem. Phys.*, **31**, 944 (1959).

56. F. E. Belles and M. R. Lauver, *J. Chem. Phys.*, **40**, 415 (1964); W. C. Gardiner, K. Morinaga, D. L. Ripley and T. Takeyama, *Sixth International Shock Tube Symposium*, Freiburg, April, 1967.

57. J. C. Camm, B. Kivel, R. L. Taylor and J. D. Teare, *J. Quant. Spectry. Radiative Transfer*, **1**, 53 (1961); D. Husain and R. G. W. Norrish, *Proc. Roy. Soc. (London)*, **A273**, 145 (1963); G. Herzberg, *Molecular Spectra and Molecular Structure: I. Spectra of Diatomic Molecules*, Van Nostrand, Princeton, N. J., 1950.

58. H. Guenebaut, G. Pannetier and P. Goudmand, *Compt. Rend.*, **251**, 1116 (1960); H. Guenebaut, G. Pannetier and P. Goudmand, *Bull. Soc. Chim. France*, **1962**, 80.

59. R. G. Bennett and F. W. Dalby, *J. Chem. Phys.*, **32**, 1716 (1960).

60. H. E. Avery, J. N. Bradley and R. Tuffnell, *Trans. Faraday Soc.*, **60**, 335 (1964).

61. M. W. P. Cann and S. W. Kash, *J. Chem. Phys.*, **41**, 3055 (1964).

62. J. N. Bradley, R. Butlin and D. Lewis, *Trans. Faraday Soc.*, **63**, 2962 (1967).

63. G. G. Mannella, *J. Chem. Phys.*, **36**, 1079 (1962); G. G. Mannella, *J. Chem. Phys.*, **37**, 678 (1962).

64. H. Guenebaut, G. Pannetier and P. Goudmand, *Compt. Rend.*, **251**, 1480 (1960).

65. K. H. Welge, *J. Chem. Phys.*, **45**, 4373 (1966).

66. R. W. Nicholls and W. H. Parkinson, *J. Chem. Phys.*, **26**, 423 (1957); R. W. Nicholls and W. H. Parkinson, *Can. J. Phys.*, **36**, 625 (1958); R. W. Nicholls, M. D. Watson and W. H. Parkinson, *J. Roy. Astron. Soc. Can.*, **53**, 223 (1959); W. H. Parkinson, *Proc. Phys. Soc.*, **78**, 705 (1961); W. R. S. Garton, W. H. Parkinson and E. M. Reeves, *Proc. Phys. Soc. (London)*, **80**, 860 (1962).

67. R. W. Nicholls, *J. Roy. Astron. Soc. Can.*, **53**, 109 (1959).

68. R. W. Nicholls and M. D. Watson, *Nature*, **188**, 568 (1960).

3

Chemistry of Electronically Excited States of Organic Molecules

A. Kearwell and F. Wilkinson

3.1 INTRODUCTION

Each electronic state of a molecule has a different electron distribution and therefore different chemical properties. Despite the fact that all molecules can exist in many possible electronic states, chemists have been preoccupied for over a century with the accumulation of knowledge about only one of these states, the lowest or ground state. This is understandable since at normal temperatures, for a system in equilibrium, the fraction of molecules in electronically excited states is very small indeed. However, electronically excited states can be produced easily and conveniently by irradiation with ultraviolet or visible light and recently there has been an increasing interest in the resulting photochemical reactions. It is difficult to overemphasise the importance of such reactions since they are responsible, directly or indirectly, for almost all our food supplies and sources of energy and may well have played an essential role in the origins of life itself.

It has always been the aim of photochemists to interpret observed photochemical behaviour in terms of the originally excited state, but in the past the lack of knowledge concerning various photophysical processes, including inter- and intra-molecular energy transfer, has made this a difficult task. The situation is rapidly improving and in recent years there has been a considerable increase in the understanding of photophysical processes and also in the interpretation of electronic absorption and emission spectra of organic molecules especially those containing π electrons. This knowledge, coupled with decisive experiments, has enabled the electronically excited states which are undergoing reaction to be assigned in a number of cases.

The emphasis of this review is directed towards a mechanistic understanding of the reactions of excited states occurring mainly in solution. This requires a spectroscopic knowledge of the initially excited state and must take into account the various photophysical processes which may result in the production of the reactive state or in the loss of energy by other modes of decay. Although it is essential to know which state is reactive and why, in order to fully understand any photochemical reaction, it must be remembered that the secondary reactions of intermediates and products are often of utmost importance in governing the overall quantum yield of reaction. Any information concerning the dependence of quantum yields of reaction on excitation wavelength, intensity, temperature, concentration, solvent nature, etc., helps with the final elucidation of the overall reaction mechanism. The detection and identification of transient intermediates and an examination of their kinetic behaviour can also prove invaluable.

At present the complete mechanisms of only a few photochemical reactions are established but this position is likely to improve since this field of research is expanding rapidly. The few representative examples discussed in the text will illustrate the kind of understanding which is possible and point the way for future developments which should eventually provide a systematic guide for synthetic work and an understanding of biological photodependent processes.

3.2 PRODUCTION OF ELECTRONICALLY EXCITED STATES

The first four methods are the subject of other articles in this volume and in this series, and the reader is referred to them for fuller treatment.

3.2.1 Thermal Excitation

Most electronically excited states lie at least 50 kcal/mole above the ground state so that the fraction of molecules produced by thermal population is very small indeed. For two electronic states separated by 50 kcal/mole at 298°K Boltzmann's distribution law gives N_u/N_l = $\exp(-E/RT) \approx 2 \times 10^{-37}$ where N_u and N_l are the number of molecules in the upper and lower levels respectively. At higher temperatures the fraction of molecules in electronically excited states increases but so does the population of highly vibrational excited ground-state molecules. This makes the assignment of any observed reaction to a particular electronically excited state difficult. That electronically excited states are produced is evident from the observed characteristic emissions in flames and shock tubes, etc., but such conditions lead to decomposition of most molecules and only in the case of atoms and very simple molecules have such studies led to any detailed knowledge concerning the properties of electronically excited species (see Chapter 1).

3.2.2 Chemical Production

In most elementary chemical reactions the reactants need to gain energy to reach the transition state and then pass over to products with energy equal to the enthalpy of activation, $\Delta H^{0\ddagger}$, minus the enthalpy of reaction, ΔH^0 (see Figure 3.1). It is possible that electronically excited products

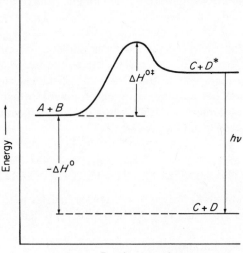

Figure 3.1 Schematic diagram showing reaction pathway for a chemiluminescent reaction
$$A + B \rightarrow C + D + h\nu$$

will result if $\Delta H^{0\ddagger} - \Delta H^0$ is greater than the electronic excitation energy of one of the products. This is only likely for exothermic reactions where ΔH^0 is of course negative. Any emission arising from electronically excited states formed in this manner is known as chemiluminescence.

However there are a very limited number of efficient chemiluminescent reactions. The most efficient are those found in biological systems, an excellent example being the firefly which is estimated to have an efficiency close to 100%, i.e. all molecules which react yield excited products which luminesce[1].

3.2.3 Electrical Discharge

When a material is subjected to an electrical discharge electronically excited states are often produced but unfortunately complex molecules are often fragmented and ionized under these conditions. So-called electrodeless discharges at radio or microwave frequencies are often used for dissociating diatomic gases in gas flow systems. The electrodes are placed on the outside of the flow tube and gases can be introduced downstream from the excitation region so that excited states may be formed as a result of atom–molecule reactions, etc.

3.2.4 Excitation with Ionizing Radiation

As the name implies α, β and γ rays are capable of ionizing the medium through which they pass. Excited states are often produced as a result of recombination reactions and the luminescence from such states is of primary importance in scintillator methods for detecting ionizing radiation (see Chapter 4). Chemical reactions, as opposed to nuclear reactions, of irradiated samples have been the subject of much study. The technique of pulse radiolysis, which is the ionizing radiation analogue of flash photolysis (see below), has been shown to be capable of producing excited states in high yields enabling interesting studies to be made[2].

3.2.5 Excitation by Absorption of Ultraviolet and Visible Light

Only the briefest of introductions to molecular electronic absorption spectra can be given here. Detailed discussions of the interaction of electromagnetic radiation with matter can be found in various books on spectroscopy[3] and quantum mechanics[4].

Electromagnetic radiation possesses properties associated with both particles and waves and the first condition which must be fulfilled if light of frequency v is to be absorbed and promote a molecule from an electronic state Ψ_l to another state Ψ_m is that the energy difference between the two states should be equal to the energy hv of the bombarding photons, i.e.

$$E_m - E_l = hv \tag{3.1}$$

The probability of a photon being absorbed in unit time is $B_{lm}\rho$ where ρ is the density of the electromagnetic radiation with the resonance frequency v and B_{lm} is Einstein's coefficient of absorption.

Absorption is mainly a result of the interaction of the oscillating electric vector of the electromagnetic radiation with the charged particles within the molecule and according to time-dependent perturbation theory,

$$B_{lm} = \frac{8\pi^3}{3h^2}|\mathbf{M}_{lm}|^2 \tag{3.2}$$

where $\mathbf{M}_{lm}$ is the transition moment which is given by

$$\mathbf{M}_{lm} = \int \Psi_l \left[\sum_i z_i e \mathbf{r}_i \right] \Psi_m \, d\tau \tag{3.3}$$

and $\mathbf{r}_i$ is the position vector of the ith particle of charge $z_i e$ in the molecule.

This integral may be zero for a variety of reasons and the corresponding transition is said to be *forbidden* as a result of certain selection rules. The integral vanishes if there is a change of electron spin upon excitation and transitions between states of different multiplicities are said to be *spin-forbidden*. In order for $\mathbf{M}_{lm}$ to be large the wave functions must have considerable spatial overlap and the symmetries of the wave functions and of $\mathbf{r}_i$ must be such that they do not give an odd function since the integral overall space of an odd function is zero. This requirement leads to certain symmetry selection rules. If the direct product of the symmetries of the two wave functions $\Psi_l\Psi_m$ does not contain a term which has the symmetry properties of $\mathbf{r}_i$, i.e. of a translatory motion, then the integral becomes zero. For molecules with a centre of symmetry all wave functions are either symmetric or antisymmetric with respect to inversion through the centre (i.e. either gerade g or ungerade u) and the integral vanishes unless the direct product of the wave functions is ungerade. The parity selection rule is often written as u ↔ g but u ↮ u and g ↮ g where the crossed arrows indicate a forbidden transition. Actually *forbidden* transitions do occur and calculations by less approximate theories show that they are expected to occur but with much smaller intensities than those termed *allowed*.

So far no mention has been made of vibrational changes which can accompany electronic transitions. These can distort the symmetry of either electronic state and a transition which is forbidden in the absence of vibrations may not be completely forbidden because of distortion by an appropriate vibration. The lowest singlet–singlet transition in benzene which occurs near 260 mμ is an example of a forbidden absorption band which becomes weakly allowed due to vibrational interaction.

The experimental measure of the transition probability is the extinction coefficient, ε, which is defined for monochromatic light by the Beer–Lambert law

$$\log_{10} \frac{I_0}{I} = \varepsilon C l \tag{3.4}$$

where I_0 and I are the intensities of the incident and transmitted light respectively, C is the concentration of the absorbing species and l is the pathlength. Electronic transitions with large transition moments have large extinction coefficients at their wavelength maxima. However, a better measure of transition probability is the oscillator strength, f, which is proportional to the area of the absorption band as follows

$$f = \frac{2303 mc^2}{\pi e^2 N} \int \varepsilon(\tilde{v}) \, d\tilde{v} \tag{3.5}$$

where m is the mass and e the charge of an electron, and $\varepsilon(\tilde{v})$ is the molar decadic extinction coefficient at wavenumber $\tilde{v}$. The oscillator strength can also be related to the transition moment

$$f = \frac{8\pi^2 mv}{3he^2} |\mathbf{M}_{lm}|^2 \tag{3.6}$$

Electromagnetic radiation can induce emission from an upper state with a probability equal to that of induced absorption from the lower state since $B_{ml} = B_{lm}$ where B_{ml} is Einstein's coefficient for induced emission. However, emission from an upper state can also take place spontaneously (i.e. in the absence of any applied electromagnetic radiation) with a probability A_{ml} which is given by

$$A_{ml} = \frac{8\pi hv^3}{c^3} B_{ml} \tag{3.7}$$

as was first shown by Einstein.

Electronic transitions usually occur at high frequencies and as A_{ml} is proportional to v^3 it follows that under most conditions, even in the presence of exciting light (except at the very high radiation densities found in and produced by lasers), emission is due to spontaneous decay. When excitation is stopped, the number of molecules remaining at time t, N_m, will be related to those present initially, N_m^0, by the equation

$$N_m = N_m^0 \exp(-A_{ml}t) \tag{3.8}$$

Thus $\tau_R = 1/A_{ml}$ is the mean radiative lifetime and if there are other states below m the radiative lifetime $\tau_R = 1/\Sigma_1 A_{ml}$. The actual lifetime of an excited state τ will be less than this unless the quantum yield of emission ϕ_E is unity.

$$\phi_E = \frac{\text{Number of quanta emitted}}{\text{Number of quanta absorbed}}$$

If the competing non-radiative processes are also first order then

$$\phi_E = \tau/\tau_R \tag{3.9}$$

3.3 DESCRIPTION OF ELECTRONICALLY EXCITED STATES OF ORGANIC COMPOUNDS

3.3.1 Singlet and Triplet States

Most molecules contain an even number of electrons. In molecular orbital theory the ground state corresponds to an arrangement where the lowest energy molecular orbitals are filled with pairs of electrons with spins opposed in accordance with Pauli's principle. This arrangement

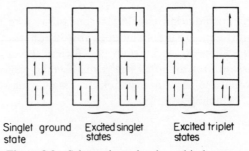

Singlet ground   Excited singlet   Excited triplet
state   states   states

Figure 3.2   Schematic molecular orbital representations of electronic states

corresponds to a singlet state since the total spin angular momentum quantum number S is zero and the multiplicity, $M = 2S + 1$, is unity. Electronically excited states can be adequately described as those in which one electron has been promoted from a lower to a higher energy molecular orbital as shown in Figure 3.2. Arrangements which have unpaired electrons with parallel spins have $S = 1$ and $M = 3$ and therefore correspond to three closely spaced energy levels, referred to collectively as a *triplet* state. Thus there are excited singlet states with corresponding triplet states of lower energy (Hund's rule) for each different electron configuration (see Figure 3.3).

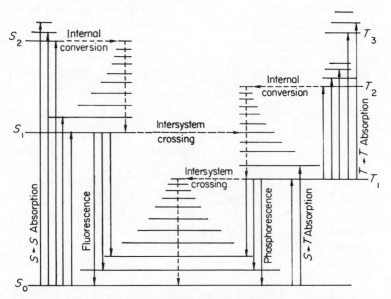

Figure 3.3 Schematic diagram showing radiative and non-radiative transitions (continuous and broken lines respectively) between singlet (S) and triplet (T) states of a typical organic molecule

3.3.2 Spin–Orbit Coupling

When one talks of singlet and triplet states one is assuming that the orbital and spin angular momenta of an electron in any molecule do not interact. However, allowance can be made for a small amount of spin–orbit coupling in molecules by defining a nominal triplet state wave function

$$\Psi_T = \Psi_T^0 + \lambda \Psi_S^0$$

where

$$\lambda = \left| \frac{\Psi_S^0 H_{so} \Psi_T^0 \, d\tau}{E_T - E_S} \right|$$

and Ψ_S^0 and Ψ_T^0 are the wave function for 'pure' singlet and 'pure' triplet states derived assuming no spin–orbit interaction. E_S and E_T are the singlet and triplet state energies and H_{so} is the Hamiltonian operator for spin–orbit perturbation. Although Ψ_T is no longer a pure triplet state it is a nominal triplet state provided $\lambda \ll 1$. Similarly singlet states have a small admixture of triplet character. This explains why spin forbidden singlet–triplet transitions do occur although they are very weak. The transition probability is proportional to the amount of spin–orbit coupling which is very dependent on the atomic numbers of the atoms in the molecule. The presence of heavy atoms either within the molecule or in

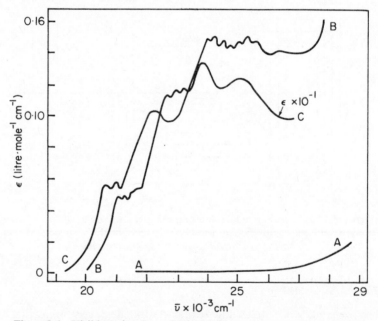

Figure 3.4 Visible region $S_0 \to T_1$ absorption curves at room temperature. Curve A, 1-chloronaphthalene; curve B, 1 volume 1-chloronaphthalene + 2 volumes ethyliodide; curve C, 1-iodonaphthalene (McGlynn, Smith and Cilento[5])

the surrounding solvent enhances the singlet–triplet transition probability (see Figure 3.4).

3.3.3 Molecular Orbital Configurations

Electronic transitions have been classified according to the orbitals which the promoted electron leaves and enters. Some of the more important transitions are $n \rightarrow \pi^*, \pi \rightarrow \pi^*, \sigma \rightarrow \pi^*, n \rightarrow \sigma^*, \sigma \rightarrow \sigma^*$, given here roughly in order of increasing energy. The states produced by these promotions can be described as (n, π^*) and (π, π^*) states, etc. The photochemistry of molecules in solution is greatly simplified by the fact that the radiationless processes from higher excited states are so efficient that the lowest excited singlet and triplet states are produced rapidly no matter which state is originally excited. Photochemical properties are therefore dependent on the properties of these lowest excited states, which for molecules excited by ultraviolet and visible radiation are usually either (n, π^*) or (π, π^*) states. The energy difference between $^1(n, \pi^*)$ and $^3(n, \pi^*)$ states usually lies in the range 2000–5000 cm^{-1}, which is a good deal smaller than the splitting of many $^1(\pi, \pi^*)$ and $^3(\pi, \pi^*)$ states which can be as high as 12,000 cm^{-1} and decreases with increasing size of the molecule. A simple model explaining these differences can be found in reference (5). It follows that a molecule with $S_1 = {}^1(n, \pi^*)$ does not necessarily have a $^3(n, \pi^*)$ state as its lowest triplet state. Even when S_1 and T_1 are (π, π^*) states they often have different symmetries since some (π, π^*) states show very little singlet–triplet splitting[6].

3.4 RADIATIVE AND NON-RADIATIVE TRANSITIONS IN ORGANIC MOLECULES

Figure 3.3 shows the intramolecular radiative and non-radiative transitions which occur in a typical organic molecule. The same information may be represented by a reaction scheme which allows a definition of some of the terms introduced in the following sections.

Process	Description	Rate
$S_0 + hv \rightarrow S_x$	singlet–singlet absorption	I_a
$S_x \rightarrow S_1$	internal conversion	$k_{S_x \rightarrow S_1}[S_x]$
$S_1 \rightarrow S_0 + hv_F$	fluorescence	$k_F[S_1]$
$S_1 \rightarrow S_0$	internal conversion	$k_{S_1 \rightarrow S_0}[S_1]$
$S_1 \rightarrow T_1$	intersystem crossing	$k_{S_1 \rightarrow T_1}[S_1]$
$T_1 \rightarrow S_0 + hv_P$	phosphorescence	$k_P[T_1]$
$T_1 \rightarrow S_0$	intersystem crossing	$k_{T_1 \rightarrow S_0}[T_1]$

where S_0 is the ground state, S_x is any vibrationally excited singlet state and S_1 and T_1 are the thermally equilibrated lowest excited singlet and triplet states respectively.

The quantum yields of fluorescence and phosphorescence will be represented as ϕ_F and ϕ_P respectively and the lifetimes of the excited singlet and triplet states by τ_{S_1} and τ_{T_1}. Equations for these quantities are derived below using stationary state conditions but these values would still be correct for other conditions, e.g. with flash excitation. Under stationary state conditions with constant light intensity

$$\phi_F = \frac{k_F[S_1]}{I_a} \quad \text{and} \quad \phi_P = \frac{k_P[T_1]}{I_a}$$

also

$$\frac{d[S_x]}{dt} = 0, \quad \frac{d[S_1]}{dt} = 0 \quad \text{and} \quad \frac{d[T_1]}{dt} = 0$$

$$[S_1] = \frac{I_a}{k_F + k_{S_1 \to S_0} + k_{S_1 \to T_1}} = I_a \tau_{S_1}$$

and

$$[T_1] = \frac{k_{S_1 \to T_1}[S_1]}{k_P + k_{T_1 \to S_0}} = k_{S_1 \to T_1} \tau_{T_1}[S_1]$$

Therefore

$$\phi_F = \frac{k_F}{k_F + k_{S_1 \to S_0} + k_{S_1 \to T_1}} \tag{3.10}$$

and

$$\phi_P = \left(\frac{k_{S_1 \to T_1}}{k_F + k_{S_1 \to S_0} + k_{S_1 \to T_1}} \right) \left(\frac{k_P}{k_P + k_{T_1 \to S_0}} \right) \tag{3.11}$$

$$= \phi_T \theta_P \tag{3.12}$$

where ϕ_T is the quantum yield of triplet state production and θ_P is the fraction of triplet molecules formed which phosphoresce.

3.4.1 Singlet–Singlet Absorption

$\pi \to \pi^*$ transitions which occur in the visible and near ultraviolet spectral regions are found in aromatic hydrocarbons and other unsaturated compounds including carbonyl compounds. Absorption varies from very strong to weak with ε max $\sim 10^4$ l mole^{-1} cm^{-1} for allowed $\pi \to \pi^*$ transitions. For unsaturated molecules which contain heteroatoms with non-bonding pairs of electrons there is a further possible transition, $n \to \pi^*$, which is, in many compounds, the lowest energy transition. It should also be noted that such transitions have much lower transition probabilities, typical values of ε max being $\sim 10^2$ l mole^{-1} cm^{-1}. Another distinguishing feature is that whereas $n \to \pi^*$ transitions show blue shifts in polar solvents, $\pi \to \pi^*$ transitions show red shifts.

3.4.2 Triplet–Triplet Absorption

Triplet states can be formed in high yields in flash photolysis experiments which make the measurement of triplet–triplet absorption spectra possible (see Figure 3.5). The transition is spin-allowed and although only a few extinction coefficients have been measured one would expect as much variety of transition probability as is found in singlet–singlet spectra. Extinction coefficients of triplet anthracene and naphthalene have been

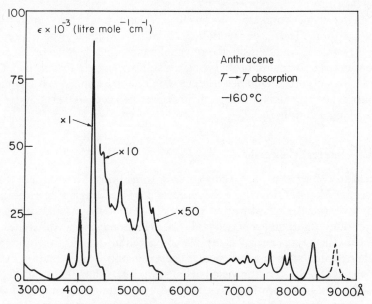

Figure 3.5   Triplet–triplet absorption spectrum of anthracene in solution (ethanol-3:methanol-1) at 113°K (Astier and Meyer[7])

measured by various authors. Methods which use direct measurement of the decrease in ground state population in order to calculate triplet state concentrations and thus triplet extinction coefficients seem to be the most reliable[7].

3.4.3 Singlet–Triplet Absorption

Since singlet–triplet absorption is spin forbidden it is very weak. From a knowledge of the radiative lifetimes of the reverse process of phosphorescence ε max values for transitions from the ground states to give $^3(\pi, \pi^*)$ or $^3(n, \pi^*)$ states are estimated to be $\sim 10^{-5}$ and $\sim 10^{-2}$ l mole^{-1} cm^{-1} respectively. This is the reverse order of transition probabilities to that found for singlet–singlet absorption and the reason for this is that first order spin–orbit coupling is forbidden between states of the same configuration. Thus (n, π^*) states couple through spin–orbit interaction with (π, π^*) states and vice versa. The admixed singlet states in a nominal $^3(n, \pi^*)$ state are therefore $^1(\pi, \pi^*)$ states with high radiative probabilities while $^3(\pi, \pi^*)$ states are admixed with $^1(n, \pi^*)$ or $^1(\sigma, \pi^*)$ states which have low transition probabilities.

Singlet–triplet transitions to produce $^3(\pi, \pi^*)$ states show much larger heavy atom effects (see Section 3.3.2) than do those which produce $^3(n, \pi^*)$ states[8]. In addition, paramagnetic molecules help to break down the spin selection rules, probably as a result of complex formation. Thus Evans has obtained enhanced singlet–triplet absorption spectra of many compounds by placing oxygen at pressures of up to 100 atm above the absorbing solutions[9].

3.4.4 Internal Conversion

Internal conversions are radiationless transitions between electronic states of like multiplicity. The process involves an isoenergetic nonradiative transition as shown in Figure 3.3. The lower excited state gains an amount of internal energy equal to the electronic energy difference. In solution this excess energy is rapidly converted into heat. Emission from higher excited states is not usually detectable and from Equation (3.9) it follows that if the quantum yield of emission is less than 0.1% the lifetime of the excited state, τ, which is governed by the rate of the non-radiative process, will be 1000 times less than τ_R. Triplet–triplet fluorescence has never been observed, showing that internal conversion between triplet states must also be very efficient. Rate constants for internal conversions between excited states are usually estimated to be $\sim 10^{12} \text{ s}^{-1}$.

In aromatic hydrocarbons and a number of other compounds it has been shown that internal conversion to the ground state is much slower than this or absent altogether[10,11]. The reasons for this are not fully understood but the large electronic energy difference between S_1 and S_0 is thought to be a major contributing cause (see Section 3.4.6).

3.4.5 Fluorescence

Emission between states of like multiplicity is termed fluorescence. Usually in solution or in the gas phase at pressures greater than a few cm of Hg, fluorescence arises only from the thermally relaxed lowest excited singlet state as shown in Figure 3.3.

The Franck–Condon principle asserts that an electronic transition takes place so quickly ($\sim 10^{-15}$ s) that the internuclear distance can be regarded as fixed during the transition. It follows that electronic transitions as shown in Figure 3.6 must be represented on a diagram by means of a vertical line. The relative intensities of the different vibrational components of any electronic transition depend on the values of ψ in the upper and lower states, i.e. they depend on overlap of the vibrational wave functions (see Figure 3.6). This determines the relative intensities of the various vibrational sub-bands but not the total integrated intensity of an electronic band. When the potential energy surface of the electronically excited state is similar to that of the ground state there is mirror image relationship between the absorption spectrum of the lowest energy transition and the fluorescence spectrum (see Figure 3.6).

The quantum yield of fluorescence is equal to the fraction of excited molecules which internally convert to the lowest excited singlet state, S_1, multiplied by the fraction of molecules in S_1 which emit. Many molecules have a quantum yield of fluorescence which is independent of wavelength (see Table 3.1) indicating that, irrespective of which state is initially excited, internal conversion to S_1 takes place with unit efficiency. The quantum yield of fluorescence is therefore determined by the relative efficiencies of emission and of non-radiative decay from S_1 (see Equation 3.10).

The radiative lifetime of the first excited singlet state may be calculated by combining Equations (3.2), (3.5), (3.6) and (3.7) to give

$$\frac{1}{\tau_R} = A_{ml} = \frac{8 \times 2303 \pi c}{N} \tilde{v}^2 \int \varepsilon(\tilde{v}) \, d\tilde{v} \qquad (3.13)$$

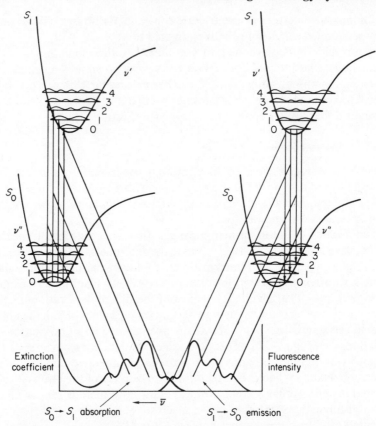

Figure 3.6 Simplified potential energy curves with vibrational probability functions showing how a mirror image relationship can arise between the electronic absorption and emission bands

However this equation is strictly applicable only to atomic systems in the gas phase where the transitions give sharp lines. Modified equations, which allow for the refractive index of the solvent and the fact that absorption and emission spectra of polyatomic molecules are spread over a range of frequencies, give good agreement with experimental values[12,13] (see Table 3.2). One such equation[12] is

$$\frac{1}{\tau_R} = \frac{8 \times 2303\pi c}{N} n^2 \langle \tilde{v}_f^{-3} \rangle_{Av}^{-1} \int \varepsilon(\tilde{v}) \, d\ln \tilde{v} \qquad (3.14)$$

Table 3.1 [a]

Substance	Solvent	Excitation wavelength range (mμ)	ϕ_F
Anthracene	Ethanol	210–310	0·30
2-Aminoanthracene	Ethanol	210–470	0·45
Fluorene	Hexane	205–310	0·54
Fluorene	Ethanol	205–310	0·54
Phenol	Water	210–300	0·22
Skatole	Water	210–300	0·42
Riboflavin	Water	210–500	0·26
N-methyl acridinium chloride	Water	210–470	1·01
Fluorescein	0·1M NaOH	210–530	0·92
Eosin	0·1M NaOH	210–590	0·19

[a] Values taken from Weber and Teale[123].

where n is the refractive index and

$$\langle \tilde{v}_f^{-3} \rangle_{Av}^{-1} = \frac{\int I(\tilde{v}) \, d\tilde{v}}{\int \tilde{v}^{-3} I(\tilde{v}) \, d\tilde{v}}$$

$I(\tilde{v})$ is the fluorescence intensity in units of relative quanta per unit frequency interval at the wavenumber $\tilde{v}$. Very rough values of expected

Table 3.2

Comparison of measured lifetimes with those calculated using the Strickler–Berg equation [a]

Substance	Solvent	ϕ_F	τ_R(calc) s $\times 10^9$	τ measured s $\times 10^9$	τ(calc) = τ_R(calc) $\times \phi_F$ s $\times 10^9$
Perylene	Benzene	0·89[b]	4·82	4·79	4·29
Acridone	Ethanol	0·83[b]	14·53	11·80	12·06
9-Aminoacridine	Ethanol	0·99[c]	15·60	13·87	15·43
9-Aminoacridine	Water	0·98[c]	16·73	16·04	16·40
Fluorescein	H_2O/NaOH	0·93[c]	4·70	4·02	4·37
Rhodamine B	Ethanol	0·97[c]	6·20	6·16	6·01

[a] Lifetime values taken from Strickler and Berg[12].
[b] Values from Melhuish[124].
[c] Values from Weber and Teale[125].

lifetimes can be calculated from the equation

$$\tau_R \approx \frac{10^{-4}}{\varepsilon_{max}} \, s \tag{3.15}$$

Thus $^1(\pi, \pi^*)$ states have radiative lifetimes as short as 10^{-8} s while τ_R for $^1(n, \pi^*)$ states is nearer 10^{-6} s. Molecules in which S_1 is a $^1(n, \pi^*)$ state are usually non-fluorescent partly because of the much longer radiative lifetime and partly because of an increase in the rate of inter-system crossing.

There are many mechanisms for the production of the fluorescent state which result in a measured lifetime for fluorescence much longer than the radiative lifetime. An example is the temperature-dependent delayed fluorescence which arises as a result of thermal excitation from the lowest triplet state[14]. This is most likely for molecules in which the energy gap between S_1 and T_1 is small. The measured lifetime of such delayed fluorescence would be equal to the lifetime of the triplet state from which S_1 is formed. This longer lifetime in no way implies that the transition probability of fluorescence has changed. Other mechanisms which give rise to delayed fluorescence are mentioned later.

3.4.6 Intersystem Crossing

Since singlet–triplet absorption is very weak the population of triplet states by direct excitation is extremely difficult. However the triplet state can be produced in high yields as a result of non-radiative processes following singlet–singlet absorption. Intersystem crossing is defined as a radiationless transition which involves a change of multiplicity. It is thus a spin forbidden internal conversion and yet intersystem crossing from S_1 competes effectively in many molecules with the spin allowed radiative and non-radiative transitions from S_1 to S_0. Intersystem crossing from T_1 to S_0 is much less efficient and in rigid glasses or plastics a typical value for the rate constant of intersystem crossing from the lowest triplet (π, π^*) state is $k_{T_1 \to S_0} \sim 10^{-1} \, s^{-1}$. Internal heavy atom effects have been examined by measuring the fluorescence and phosphorescence yields and the phosphorescence decay of naphthalene and its 1-halo derivatives in rigid media (see Table 3.3). Heavy atoms are found to increase the rate of intersystem crossing from $^3(\pi, \pi^*)$ states but not from $^3(n, \pi^*)$ states[15].

It has been suggested earlier that the larger the electronic energy difference between two states the less likely are radiationless transitions. There are no electronic states lying between T_1 and S_0 and therefore the

Table 3.3[a]

Heavy atom effects on fluorescence and phosphorescence yields and
phosphorescence lifetimes in alcohol/ether at 77°K

Compound	ϕ_F	ϕ_P	τ_R (s)
Naphthalene	0·29	0·03	63
1-Chloronaphthalene	0·03	0·16	1·7
1-Bromonaphthalene	0·00	0·14	0·14
1-Iodonaphthalene	0·00	0·20	0·01

[a] Values from Ermolaev[15].

transition must be a direct one. Although the experimental points are
somewhat scattered the values of $k_{T_1 \to S_0}$ do decrease with increasing size
of the T_1–S_0 energy gap (see Figure 3.7). The theory of radiationless

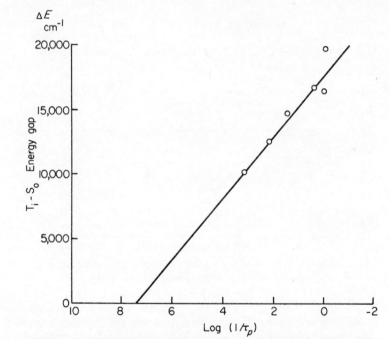

Figure 3.7 Triplet decay constants for several perprotonated hydrocarbons plotted as a function of the T_1–S_0 energy gap. The data were
taken for the perprotonated molecules at room temperature in poly-
methylmethacrylate (Kellogg and Wyeth[141])

transitions developed by Robinson and Frosch[16], explains this effect in terms of Franck–Condon factors. The initial state of the molecule is the lowest vibrational level of T_1 and the final state can be any of a number of degenerate highly vibrational levels of the ground state. The probability of a radiationless transition is proportional to the square of the vibrational overlap integral between the initial and final states. When the energy gap is large there are many degenerate vibrational levels of the ground state to which the molecule can possibly convert and the process might be expected to occur with high efficiency. However vibrational wave functions with high vibrational quantum numbers possess a large number of nodes and this leads to a reduction in the amount of overlap, as shown diagrammatically in Figure 3.8. This effect more than compensates for the

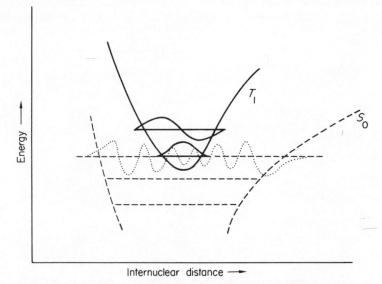

Figure 3.8   Diagram showing overlap of vibrational wave functions
involved in intersystem crossing from $T_1 \rightarrow S_0$

greater number of levels. C—H vibrations are the highest frequency vibrations found in aromatic hydrocarbons and vibrational levels of S_0, which include harmonics and combinations of C–H vibrations, will have the largest overlap integrals since the wave function will have a less oscillatory nature. When C—H bonds are replaced by C—D the degenerate vibrational levels will have higher vibrational quantum numbers and the radiationless transition probability will be much less. This leads to an increase in the measured phosphorescence lifetime when aromatic hydrocarbons are deuterated[17] (see Table 3.4).

Table 3.4

Comparison of phosphorescence lifetimes of perprotonated and perdeuterated aromatic compounds

Compound	$\tau_{T_1}(s)$ (perprotonated)	$\tau_{T_1}(s)$ (perdeuterated)	Solvent	Ref.
Benzene	16	26	Argon (4·2°K)	17
Naphthalene	2·25	18·3	EPA (77°K)	126
Phenanthrene	3·68	15·2	EPA (77°K)	126
Pyrene	0·5	3·2	EPA (77°K)	127
Chrysene	2·5	13	Solid Matrix (77°K)	128,129
Triphenylene	14·8	22·1	EPA (77°K)	126
Biphenyl	4·2	10·3	EPA (77°K)	127
p-Terphenyl	2·6	5·3	EPA (77°K)	127

The rate constants for intersystem crossing from S_1 have values of $k_{S_1 \to T_1} \sim 10^8$ s^{-1} when S_1 is a $^1(\pi, \pi^*)$ state and they may be as high as 10^{10} s^{-1} when S_1 is a $^1(n, \pi^*)$ state[18]. These much higher values of $k_{S_1 \to T_1}$ compared with those of $k_{T_1 \to S_0}$ are almost certainly due to the fact that intersystem crossing from S_1 takes place to triplet states lying between S_1 and T_1 so that the electronic energy difference between the electronic states involved in the transition is very small.

3.4.7 Phosphorescence

Phosphorence was originally defined as a long-lived emission but this definition has been abandoned by photochemists who reserve the term for emission between states of different multiplicities. Usually phosphorescence occurs with high efficiency only in rigid media although a few compounds, such as biacetyl, emit in all three phases. As with excited singlet states, only the lowest triplet state is found to emit. The process is spin forbidden and therefore long-lived with radiative lifetimes of $\sim 10^{-2}$ s for $^3(n, \pi^*)$ states which usually have high quantum yields of phosphorescence (see Table 3.5) while $^3(\pi, \pi^*)$ states may have lifetimes in excess of 10 s.

Table 3.5 [a]

Absolute quantum yields of fluorescence and phosphorescence and phosphorescence lifetimes of aromatic compounds in alcohol ether at 77°K

Compound	ϕ_F	ϕ_P	$\tau_R(s)$
Benzophenone	0·00	0·74	$6\cdot2 \times 10^{-3}$
Benzaldehyde	0·00	0·49	$2\cdot9 \times 10^{-3}$
Acetophenone	0·00	0·62	$3\cdot6 \times 10^{-3}$
m-Iodobenzaldehyde	0·00	0·64	—
Phenanthrene	0·12	0·135	22
Naphthalene	0·29	0·03	63
Quinoline	0·053	0·10	13
1-Methylnaphthalene	0·43	0·023	50

[a] Values from Ermolaev[15].

Phosphorescence spectra quite often show good mirror image relationships about their 0, 0 bands with the corresponding singlet–triplet absorption spectra. The 0, 0 transitions found from phosphorescence spectra compare very well with the values obtained from oxygen perturbed measurements of singlet–triplet absorption[9]. Equation (3.12) shows that the phosphorescence yield is equal to the product of the quantum yield of triplet state formation, which for many hydrocarbons[19] is $1 - \phi_F$ and for many carbonyl compounds is unity[10], multiplied by the fraction of triplet molecules which emit. Various phosphorescence yields and lifetimes are given in Table 3.5.

The presence of heavy atoms increases the quantum yield of triplet state formation, ϕ_T, and decreases the measured phosphorescence lifetime due to an increase in spin–orbit coupling (see Table 3.3).

3.5 PROPERTIES OF ELECTRONICALLY EXCITED STATES

3.5.1 Excited State Dipole Moments

The electric dipole moment of a molecule is a physical quantity determined by the geometrical arrangement of nuclei and the distribution of electrons around them. Since the absorption of ultraviolet or visible light leads to a redistribution of electrons it follows that the dipole moment of a molecule in an excited state will be different from that of the ground state. In the case of fairly small molecules electronic excitation may lead to a change in the distribution of nuclei in the equilibrium excited state. Formaldehyde is planar in the ground state but pyramidal in the excited state, corresponding roughly to a change in carbon hybridization from sp^2 to sp^3. Radicals such as CH_2 and NH_2 become linear in the excited state. However for large molecules such as condensed aromatic hydrocarbons it is reasonable to assume that excitation involves little change in nuclear configuration.

Ground state dipole moments may be determined from capacitance measurements or from the Stark splitting of rotational transitions in the microwave spectrum but neither of these methods can be used to measure the dipole moment of a short-lived electronically excited state. Possible ways of evaluating excited state dipole moments are outlined below and representative values are compared with ground state dipole moments.

Czekalla[20] found that if a solution of a fluorescent molecule in a non-polar solvent is irradiated in an electric field of about $100\,kV\,cm^{-1}$ the fluorescence emission has a greater degree of polarization than in the absence of the field. The increase in polarization depends on the magnitude of the excited state dipole moment, and if it can be assumed that an equilibrium alignment of molecules with the field is achieved in a time less than the fluorescence lifetime, values of the excited state dipole moment, μ_e, can be calculated. Table 3.6 gives some values obtained with benzene as solvent.

Table 3.6

Dipole moments of molecules in their ground states μ_g and in their first excited singlet states μ_e

Compound	μ_g	μ_e (Lippert[21])	μ_e (Czekalla[20])
4-Dimethylamino 4'-nitrostilbene	7·6	32	21·9
4-Dimethylamino 4'-cyanostilbene	6·1	29	>12
2-Amino 7-nitrofluorene	~7	25	20·3
4-Amino 4'-nitrodiphenyl	6·4	18	19·7

A number of workers[21,22,23,24] have attempted to obtain a satisfactory quantitative relationship between excited state dipole moments and shifts of absorption and fluorescence spectra in solution compared with gas phase values. A qualitative account of solvent shifts of absorption spectra is given by Bayliss and McRae[25]. They point out that all solution spectra are shifted to the red due to the polarization of the solvent by the transition dipole. This polarization is entirely responsible for shifts in the spectra of non-polar molecules in polar and non-polar solvents, and depends only on the solvent refractive index. When the solute has a dipole moment the polarization red shift is usually obscured by the effects of dipole–dipole or dipole-induced dipole interactions. Overall absorption shifts in solution are usually to the red if the permanent dipole moment of the solute increases during the transition and to the blue if it decreases.

Dipole-induced dipole interactions stabilize the ground and excited states of polar molecules dissolved in non-polar solvents. If the dipole moment in the excited state is greater than in the ground state then the dipole-induced dipole contribution to the solvent stabilization energy is greater in the excited state than in the ground state. The converse is true if the dipole moment of the molecule decreases on excitation and there will probably be an overall blue shift of the absorption spectrum since the polarization red shift is usually rather small. Because the solvent is non-polar, no rearrangement of solvent molecules about the solute occurs in the excited state and so the equilibrium excited state is no more stabilized than the Franck–Condon state. Similarly the Franck–Condon ground state is not destabilized with respect to the equilibrium ground state and the shift of the fluorescence spectrum will be in the same direction as the absorption spectrum (see Figure 3.9).

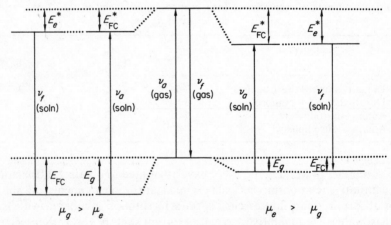

Figure 3.9 Illustrative diagram showing solvent shifts of the 0, 0 transitions of polar solutes in non-polar solvents

If both solute and solvent are polar, stabilization of the different energy states will be due predominantly to dipole–dipole forces. Following excitation the Franck–Condon excited state will be in the solvent cage appropriate to the ground state and the nett effect on the Franck–Condon excited state will depend primarily on the magnitude and direction of the dipole moment change on excitation. Because the solvent molecules have time to reorientate themselves before fluorescence emission occurs the equilibrium excited state is more stabilized than the Franck–Condon excited state. Emission occurs to give the Franck–Condon ground state which is destabilized with respect to the equilibrium ground state so that when $\mu_e > \mu_g$ the fluorescence spectrum always exhibits a red shift while the absorption spectrum may be shifted either way. If however $\mu_g > \mu_e$ the absorption spectrum will exhibit a blue shift as stated above while the fluorescence spectrum may be shifted either way (see Figure 3.10).

The solvent shift of the 0, 0 absorption band of a substance in solution and in the gas phase is equal to the difference between the solvent stabilization energies of the equilibrium ground and Franck–Condon excited states.

$$\Delta\tilde{v}_{abs} = \tilde{v}_{abs}\,(\text{soln}) - \tilde{v}_{abs}\,(\text{gas}) = (E_g - E_{FC}^*)/hc$$

while

$$\Delta\tilde{v}_F = \tilde{v}_F\,(\text{gas}) - \tilde{v}_F\,(\text{soln}) = (E_e{}^* - E_{FC})/hc$$

where E_g is equilibrium ground state stabilization energy, $E_e{}^*$ equilibrium excited state stabilization energy, and E_{FC} and E_{FC}^* are the stabilization energies of the Franck–Condon ground and excited states respectively.

For the 0, 0 band $\tilde{v}_{abs}(\text{gas}) = \tilde{v}_F\,(\text{gas})$ and therefore the frequency interval between the absorption and fluorescence 0, 0 band,

$$\tilde{v}_{abs}\,(\text{soln}) - \tilde{v}_F\,(\text{soln}) = \Delta\tilde{v}_{abs} + \Delta\tilde{v}_F$$

Therefore

$$\tilde{v}_{abs}\,(\text{soln}) - \tilde{v}_F\,(\text{soln}) = [(E_g + E_e{}^*) - (E_{FC}^* + E_{FC})]/hc$$

The stabilization energies of the various states can be expressed in terms of ground and excited state dipole moments, solvent refractive index n, dielectric constant ε and the Onsager[26] radius a. In this way Lippert[21] derived the following equation

$$\tilde{v}_{abs}\,(\text{soln}) - \tilde{v}_F\,(\text{soln}) = \frac{(\mu_e - \mu_g)^2}{hca^3}\left[\frac{2(\varepsilon - 1)}{2\varepsilon + 1} - \frac{2(n^2 - 1)}{2n^2 + 1}\right] \qquad (3.16)$$

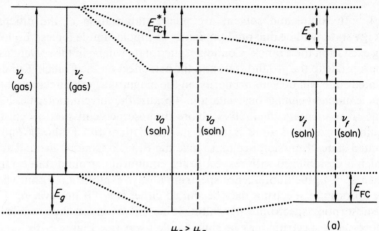

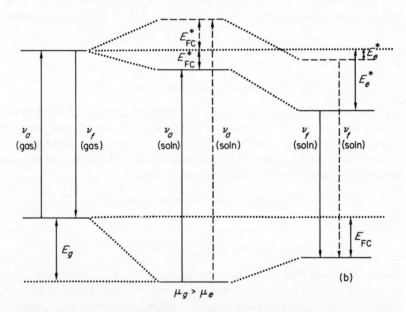

Figure 3.10   Illustrative diagrams showing solvent shifts of the 0, 0 transitions of polar solutes in polar solvents when (a) the dipole moment increases (b) the dipole moment decreases upon excitation

Plots of the difference between absorption and fluorescence maxima for a given compound in a series of solvents against the term in square brackets are linear. Since 0, 0 bands are generally not easy to observe, the assumption is made that the difference between absorption and fluorescence maxima equals $\Delta\tilde{v}_{abs} + \Delta\tilde{v}_F + a$ constant. From the slope of this type of plot, Lippert was able to calculate excited state dipole moments and his values, given in Table 3.6, may be compared with those of Czekalla.

Ledger and Suppan[27] have evaluated changes in the dipole moments of a number of molecules following excitation, using an equation which takes into account only absorption shifts as a function of the solvent dielectric constant. Their value of $\mu_e = 16.3 \, D$ for the S_1 state of 4-nitro-aniline compares quite well with the value of $14 \, D$ obtained from fluorescence polarization. The ground state dipole moment of this molecule is $6.2 \, D$ and this illustrates the considerable charge transfer which occurs in the excited state of some molecules as do the examples given in Table 3.6.

The third method of determining excited state dipole moments is from analysis of the Stark splitting of the rotational fine structure of an electronic transition[28]. The perturbation induced by the applied electric field depends on both the ground state and excited state dipole moments. Unfortunately the requirement of discrete rotational fine structure in the ultraviolet spectrum limits the method to diatomic and simple polyatomic molecules in the gas phase.

Klemperer and coworkers have applied the method to formaldehyde[29], propynal[30] and formyl fluoride[31]. The $^1(n, \pi^*)$ states of these molecules are all near-symmetric tops, consequently the rotational fine structure closely resembles that of a symmetric top and the Stark energy is linearly dependent on the applied field and the components of the excited state dipole moment along the near-symmetric top axis $\mu_e(a)$.

In the excited state of formaldehyde only the component of μ_e along the a axis is non-vanishing, but in propynal and formyl fluoride, because of the lower symmetries of the excited states of these molecules, μ_e has components along the axis a and along axis b which is in plane and perpendicular to axis a. The determination of both components of μ_e requires the measurement of second order Stark effects which would necessitate the use of extremely high applied fields. The components of μ_g and μ_e along the near-symmetric top axis a are given in Table 3.7 for each of the three molecules.

Table 3.7
Excited state dipole moments determined from
the Stark splitting or rotational fine structure of
electronic transitions

Compound	$\mu_g(a)$	$\mu_e(a)$
Formaldehyde	2·34	1·56
Propynal	2·39	0·7
Formyl fluoride	0·595	1·66

The decrease in dipole moment upon excitation for formaldehyde and propynal is consistent with a simple molecular orbital model of an $n \rightarrow \pi^*$ transition where a non-bonding electron in a p_y orbital centred on an oxygen atom is delocalized by promotion to an anti-bonding combination of carbon and oxygen p_z orbitals. However, the change in dipole moment for formaldehyde is considerably smaller than is predicted from this simple picture. One explanation is that the ground state dipole moment is reduced due to mixing of hydrogen $1s$ character with the oxygen p_y orbital. This has the effect of moving the non-bonding electrons away from the oxygen atom and towards the carbon atom thereby reducing the ground state dipole moment[31].

3.5.2 Acidity Constants of Excited States

A study of the chemical properties of a molecule in an excited state is limited to those reactions which can occur within the lifetime of the state. These reactions need not be exclusive to the excited state but the large increase in energy and the change in electron distribution should influence the reaction rates and also the positions of equilibria compared with the corresponding ground state reactions.

Acid dissociation occurs both in the ground and electronically excited levels and is amenable to fluorometric analysis. In 1931 Weber[32] noted

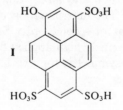

that the colour of the fluorescence of 1-naphthylamine 4-sulphonic acid depended on the pH of the solution in a region where there was no effect on the absorption spectrum. Förster[33] found a similar effect on the fluorescence of 3-hydroxypyrene 5,8,10-trisulphonic acid (I), and noted that fluorescence characteristic of the phenolate anion occurred in solutions of pH 0–2 which were too acidic for appreciable dissociation of the ground state to be expected. Changes in the ultraviolet absorption spectrum of I due to dissociation of the hydroxyl group occur in the region of pH 7 where the fluorescence spectrum shows little change. Förster concluded that acid-base equilibrium is established during the lifetime of the first excited singlet state and that for this molecule dissociation is favoured in the higher energy state.

Acid Dissociation in the First Excited Singlet State

Excited singlet state pK_a values, $pK_a^*(S_1)$, can be determined experimentally from a study of the pH dependence of the fluorescence of the acid and/or its conjugate base[34] (see Figure 3.11). Provided equilibrium

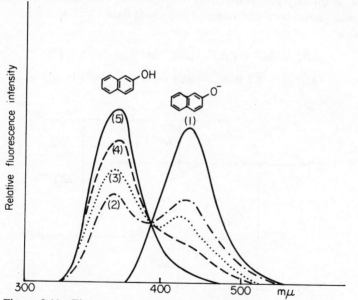

Figure 3.11 Fluorescence spectrum of 2-naphthol in solutions of different pH (from ref. 34). (1) 0·02 M NaOH, (2) 0·02 M sodium acetate + 0·02 M acetic acid, (3) pH 5–6, (4) 0·004 M HClO$_4$, (5) 0·15 M HClO$_4$

is established during the lifetime of the excited singlet state, $pK_a^*(S_1)$ will equal the pH value at which the fluorescence due to the acidic form drops to one-half the intensity observed at low pH values where it alone emits. If fluorescence from both forms can be obtained this value should coincide with the pH at which the basic form reaches one-half of its maximum fluorescence intensity.

Förster and Weller[35] devised the thermodynamic scheme illustrated in Figure 3.11 from which pK_a values of excited states can be evaluated. This method is particularly useful if equilibrium is not reached during the lifetime of the excited state when the previous method is impracticable. ΔE_{HA} and ΔE_{A^-} are the electronic transition energies from ground to first excited singlet states of the acid and the conjugate base respectively and ΔH^{0*} and ΔH^0 are the enthalpies of dissociation of the excited and ground state acids. As can be seen from Figure 3.12

$$\Delta H^0 - \Delta H^{0*} = \Delta E_{HA} - \Delta E_{A^-}$$

i.e. the difference between the heats of dissociation in ground and excited states is directly proportional to the frequency interval, $\Delta \bar{\nu}$, between the $0, 0$ absorption bands of acid and conjugate base.

Now

$$\Delta H^0 = \Delta G^0 + T\,\Delta S^0 \quad \text{and} \quad \Delta H^{0*} = \Delta G^{0*} + T\,\Delta S^{0*}$$

$$\Delta G^0 = -RT \ln K_a \quad \text{and} \quad \Delta G^{0*} = -RT \ln K_a^*$$

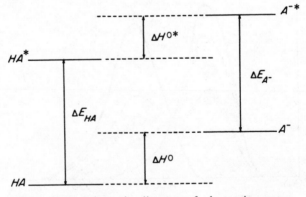

Figure 3.12 Schematic diagram of electronic energy levels of HA and A^- in the ground state and in an excited state

If it is assumed that $\Delta S^0 = \Delta S^{0*}$, then

$$\Delta E_{HA} - \Delta E_{A^-} = -RT \ln \frac{K_a}{K_a{}^*}$$

from which it follows that

$$pK_a(S_0) - pK_a{}^*(S_1) = \frac{\Delta E_{HA} - \Delta E_{A^-}}{2 \cdot 303 RT} = \frac{hc \, \Delta \tilde{\nu}}{2 \cdot 303 kT} \qquad (3.17)$$

However, $pK_a{}^*(S_1)$ values obtained by this method do not compare well with those evaluated by the first method. The reason for this is that absorption leads to a Franck–Condon excited state in a ground state solvent cage. Reorientation of the solvent molecules occurs subsequently and accompanies internal conversion to produce a thermally equilibrated excited state in equilibrium with its environment. The solvent reorientation is not fast enough to affect the absorption frequencies and the $pK_a{}^*$ values calculated from absorption spectra are therefore inaccurate. Similarly, fluorescence leads to Franck–Condon ground state molecules.

Table 3.8

Compound	$pK_a(S_0)$	$pK_a{}^*(S_1)^a$	$pK_a{}^*(S_1)^b$
2-Naphthol	9·5	3·2	3·4
p-Cresol	10·3	4·6	4·3
p-Bromophenol	9·4	3·0	3·1
m-Methoxyphenol	9·7	4·8	4·6

[a] Values obtained from relative fluorescence intensity measurements.
[b] Values obtained by using Förster cycle.

An average of the frequencies of the absorption and fluorescence maxima should provide at least partial cancellation of the solvent relaxation effects, and will approximate to the 0, 0 frequencies of the acid and conjugate base which should be used to be consistent with the thermodynamic cycle. $pK_a{}^*(S_1)$ values calculated by putting $\Delta E = hc(\tilde{\nu}_{abs} + \tilde{\nu}_F)/2$ are comparable with those obtained by measuring relative fluorescence intensities of acid and conjugate base[36] as shown in Table 3.8.

In common with aromatic hydroxy compounds, aromatic amines are also more acidic in their excited states, i.e., the equilibrium is displaced

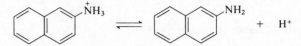

towards the right hand side of the equation when the molecules are excited. This behaviour can be correlated with a migration of electron density from the functional groups into the aromatic nucleus upon excitation[37]. Conversely both aromatic carboxylic acids and heterocyclic bases are less acidic in their excited states since charge migrates towards the substituents upon excitation[38] (see Table 3.9).

Table 3.9

Compound	$pK_a(S_0)$	$pK_a^*(S_1)$	Reference
2-Naphthylammonium cation	4·1	−2	43
1-Naphthoic acidium cation	−6·9	1·5	130
1-Naphthoic acid	3·7	10–12	43
Acridinium cation	5·45	10·3	131

In a molecule containing both acidic and basic functional groups, e.g. 3-amino-2-naphthol, the absorption spectrum characterizes three different structures over the pH range −1 to 14.

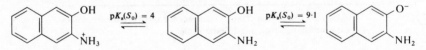

Fluorescence measurements[39] show that a fourth structure is present in the excited state, and this is identified with the zwitterion

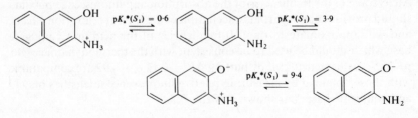

Other workers have investigated ground and excited state equilibria for hydroxyquinolines[40] and aromatic hydroxycarboxylic acids[41] which show similar behaviour.

The proton transfers discussed above have been to or from the solvent. However, in orthohydroxy carboxylic acids, e.g. salicylic acid, intramolecular proton transfer can occur in the excited state in non-protonating organic solvents because OH is more acidic and CO_2H more basic in the excited state[42].

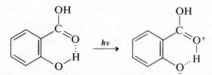

This proton transfer gives rise to exceptionally large frequency intervals between absorption and fluorescence spectra (see Section 3.4.1) which may be compared with the normal frequency shifts in o-methoxy methyl benzoate, o-methoxy benzoic acid and with those obtained when salicylic acid is dissolved in a strongly acidic or a strongly basic medium since forms (II) and (III) cannot participate in proton transfer in the excited state and normal frequency shifts are observed (see Table 3.10).

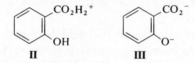

II   III

Acid-Dissociation in the Lowest Triplet State

Jackson and Porter[43] have determined pK_a values of the lowest triplet state, $pK_a^*(T_1)$, for a number of molecules from both triplet absorption

Table 3.10

Compound	Solvent	$\tilde{v}_{abs}$ (max) cm^{-1}	$\tilde{v}_{fluor}$ (max) cm^{-1}	$\Delta\tilde{v}$ cm^{-1}
Salicylic acid	Methanol	33,000	22,100	10,900
o-Hydroxy methyl benzoate	Methanol	32,800	22,300	10,500
o-Methoxy benzoic acid	Methanol	34,200	28,300	5,900
o-Methoxy methyl benzoate	Methanol	34,100	29,300	4,800
Salicylic acid	6M KOH	31,100	25,500	5,600
Salicylic acid	conc. H_2SO_4	30,400	24,700	5,700

spectra and phosphorescence spectra of the acids and their conjugate bases. In the first method the triplet states of the molecules studied were populated by intersystem crossing from the singlet manifold and triplet absorption spectra were observed using the technique of flash photolysis. The intensity of absorption of the triplet spectra of the acid and conjugate base were determined separately. At intermediate pH values where the triplet spectra of acid and conjugate base were partially superimposed it was necessary, in order to be able to compute the ratio of their concentrations, to make some assumption about the extinction coefficients of the two forms and it was assumed that they were identical at their respective band maxima. These ratios were plotted against pH and the value of the pH at which the above ratio was unity is equal to $pK_a^*(T_1)$.

$pK_a^*(T_1)$ values were also determined from phosphorescence spectra using the Förster thermodynamic cycle. The phosphorescence spectra of the acids and their conjugate bases in rigid glasses at 77°K were recorded using a conventional phosphoroscope. Frequency intervals were measured from shifts in band maxima and from the short wavelength limits (corresponding to the 0, 0 transitions). The values obtained agree well with those from triplet absorption studies despite the fact that phosphorescence measurements are made in a rigid glass in which it is doubtful whether any solvent reorientation can take place. Values are compared in Table 3.11.

From these results it is obvious that acid dissociation constants in the triplet state are much closer to ground state constants than to those of the first excited singlet state. This suggests that the electron distribution in the triplet state is similar to that in the ground state and solvent reorientation factors would therefore be expected to be small. The lowest triplet states of the molecules listed in Table 3.11 correspond closely to the $^3(\pi, \pi^*)$

Table 3.11

Compound	$pK_a(S_0)$	$pK_a^*(S_1)$	$pK_a^*(T_1)^a$	$pK_a^*(T_1)^b$
2-Naphthol	9·5	3·1	8·1	7·7
2-Naphthoic acid	4·2	10–12	4·0	4·2
1-Naphthoic acid	3·7	10–12	3·8	4·6
2-Naphthyl ammonium cation	4·1	−2	3·3	3·1
Acridinium cation	5·5	10·6	5·6	—

[a] Values obtained from relative triplet absorption measurement[43].
[b] Values obtained from phosphorescence measurements using Förster cycle[43].

states of their hydrocarbon analogues and it is therefore not very surprising that these states have $pK_a^*(T_1)$ values similar to those of the ground state.

3.5.3 Electronic Energy Transfer

Transfer of electronic excitation energy from a donor molecule, D, to an acceptor molecule, A, may be represented as

$$D^* + A \rightarrow A^* + D$$

where an asterisk denotes an electronically excited state. This process, which results in quenching of emission or reactions of D^* and sensitized emission or reactions of A^*, may take place as a result of radiative or non-radiative processes. The efficiency of radiative transfer depends on the extent of the overlap of the donor emission spectrum with the acceptor absorption spectrum. This type of transfer is simply an example of excitation by ultraviolet or visible light and is governed by the Beer–Lambert law (Equation 3.4). It depends on the path length travelled by the donor emission and therefore on the volume of the reaction vessel.

Several theories have been developed in order to explain why energy absorbed by one molecule is transferred non-radiatively to a second acceptor molecule of the same or different kind and in all cases the interaction responsible for transfer requires close resonance between the initial and final states. This is possible if vibrations are included even when there is an electronic energy difference between the excited states of the donor and acceptor as shown in Figure 3.13. Non-radiative transfer is due either to coulombic or exchange interaction. Three main ways in which energy can migrate in fluid media may be distinguished.

Long Range Single Step Transfer

Energy transfer can take place in a single step between molecules separated by distances much greater than normal collisional diameters as a result of coulombic interaction. The equations for coulombic interaction may be expanded as a multiple series and the dipole–dipole term, i.e. the term which accounts for interaction between the transition dipoles of the donor and acceptor molecules is usually predominant.

Förster[44] has given the process a quantum–mechanical treatment considering only dipole–dipole interactions and has obtained the following equation for the rate constant for transfer $k_{D^* \rightarrow A^*}$

$$k_{D^* \rightarrow A^*} = \frac{9000(\ln 10)\kappa^2 \phi_D}{128\pi^5 n^4 N \tau_D R^6} \int_0^\infty f_D(\tilde{v}) \varepsilon_A(\tilde{v}) \frac{d\tilde{v}}{\tilde{v}^4} \tag{3.18}$$

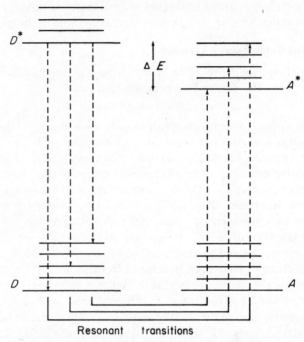

Resonant   transitions

Figure 3.13 Energy-level diagram for a donor D and
acceptor A with an electronic energy difference ΔE

where $\varepsilon_A(\tilde{v})$ is the molar decadic extinction coefficient of the acceptor at
wave number $\tilde{v}$, $f_D(\tilde{v})$ the spectral distribution of the fluorescence of the
donor (measured in quanta and normalized to unity on a wave number
scale), τ_D is the mean lifetime of the excited state, ϕ_D the quantum yield of
fluorescence of the donor, R is the distance between the molecules and
κ is an orientative factor which for a random distribution equals $(\frac{2}{3})^{1/2}$.

R_0, the distance at which transfer and spontaneous decay of the excited
donor are equally probable, is given by

$$R_0{}^6 \approx \frac{9000(\ln 10)\kappa^2 \phi_D}{128\pi^5 n^4 N \tilde{v}^4} \int_0^\infty f_D(\tilde{v})\varepsilon_A(\tilde{v})\,d\tilde{v} \tag{3.19}$$

When the transitions in both the donor and the acceptor are fully allowed
and there is good overlap between the emission spectrum of the donor and
the absorption spectrum of the acceptor, R_0 values of 50–100 Å are

predicted from this equation, and the rate constants for energy transfer due to coulombic interactions are much greater than those calculated for diffusion-controlled reactions (see below).

The spin selection rules for energy transfer by dipole–dipole interaction are very restrictive. No change of spin in either the donor or the acceptor is allowed, i.e. the multiplicities M_{D*} must equal M_D and M_A equal M_{A*}. Thus processes such as

$$D*(S_1) + A(S_0) \rightarrow D(S_0) + A*(S_1)$$

and

$$D*(S_1) + A*(T_1) \rightarrow D(S_0) + A*(T_X)$$

are fully allowed. The experimental values of R_0 for such processes give striking agreement with these theoretical predictions (see Table 3.12).

Several workers have studied intramolecular energy transfer between donor and acceptor groups which are separated by means of unsaturated

Table 3.12

Long range energy transfer distances

Donor	Acceptor	Solvent	R_0 (Å) (calc.)	R_0 (Å) (exp.)	Reference
Anthracene (S_1)	Perylene (S_0)	A	31	54	132
Perylene (S_1)	Rubrene (S_0)	A	38	65	132
9,10-Dichloro-anthracene (S_1)	Perylene (S_0)	A	40	67	132
Anthracene (S_1)	Rubrene (S_0)	A	23	39	132
9,10-Dichloro-anthracene (S_1)	Rubrene (S_0)	A	32	49	132
Phenanthrene d_{10} (T_1)	Rhodamine B (S_0)	B	45	47	133
Phenanthrene d_{10} (T_1)	Phenanthrene d_{10} (T_1)	B	40	35	70
p-Phenyl benzaldehyde (T_1)	Chrysoidin (S_0)	C	32	33	134
Triphenylamine (T_1)	Chrysoidin (S_0)	C	34	52	134
Triphenylamine (T_1)	Fuchsin (S_0)	C	29	37	134

Solvents: A, benzene at room temperature
B, cellulose acetate at 77°K
C, ethanol or dibutyl ether at 77°K

groups, e.g. Latt, Cheung and Blout[45] have measured the efficiency of energy transfer between aromatic groups attached to the hydroxy groups of a bisteroid molecule and obtained very good confirmation of Equation (3.19). By using ionic dyes with long-chain paraffin substituents arranged in monomolecular layers Kuhn and coworkers[46] have been able to build up layers of donor and acceptor chromophores at variable distances. The efficiency of transfer between donor and acceptor chromophore layers 50–100 Å apart was observed and the measurements were in very good agreement with Förster's theory.

If there is appreciable overlap of the absorption and emission spectrum of any compound, long range single step transfer between molecules of the same kind may result. This effect has been observed by using excitation with polarized light and by measuring the extent of polarization of the fluorescence. Energy transfer to molecules with different orientations leads to depolarization of the fluorescence[47] and this effect is seen to increase with concentration becoming appreciable with typical dyes when the mean distance between molecules is ~ 70 Å.

Long range transfer from the triplet state of a donor to the singlet of an acceptor was also predicted by Förster and has recently been observed[48]. In rigid media the forbidden nature of the transition in the donor results in a corresponding increase in the lifetime of the excited state of the donor. The probability of forbidden energy transfer processes such as

$$D^*(T_1) + A(S_0) \rightarrow D(S_0) + A^*(S_1)$$

and

$$D^*(T_1) + A^*(T_1) \rightarrow D(S_0) + A^*(T_X)$$

relative to spontaneous decay of the triplet state can still be high since the process of phosphorescence

$$D^*(T_1) \rightarrow D(S_0) + h\nu_p$$

is of course also spin forbidden. Transfer takes place over equally large distances (see Table 3.12) but the transfer rate constants are much lower[49].

Collisional Transfer

When donor and acceptor molecules are very close together exchange interaction will occur. Dexter[50] has given the following equation for the rate of energy transfer due to exchange interaction

$$k_{D^* \to A^*} = \frac{h}{2\pi} P^2 \, e^{-2R/L} \int f_D(\tilde{\nu}) f_A(\tilde{\nu}) \, d\tilde{\nu} \tag{3.20}$$

where $f_D(\tilde{\nu})$ and $f_A(\tilde{\nu})$ represent the donor emission and the acceptor absorption spectra respectively, normalized so that $\int f_D(\tilde{\nu}) \, d\tilde{\nu}$ and $\int f_A(\tilde{\nu}) \, d\tilde{\nu}$ both equal unity. The transfer rates are therefore independent of the oscillator strengths of both transitions. P and L are constants which are not easily related to experimentally determinable quantities. However, this equation does show that there is an expected exponential dependence on the intermolecular distance for transfer by an exchange mechanism.

Transfer processes due to exchange interaction are subject to Wigner's spin rule[51]. If S_a and S_b are the initial spin quantum numbers of the participating molecules, the resultant spin quantum number of the two species taken together must have one of the values

$$S_a + S_b, S_a + S_b - 1, S_a + S_b - 2, \ldots |S_a - S_b|$$

It follows that the spin quantum numbers of the resulting species can only have values S_c and S_d, if at least one of the values

$$S_c + S_d, S_c + S_d - 1, S_c + S_d - 2, \ldots |S_c - S_d|$$

is common to the series above. Thus processes such as

$$D^*(S_1) + A(S_0) \to D(S_0) + A^*(S_1)$$

and

$$D^*(T_1) + A(S_0) \to D(S_0) + A^*(T_1)$$

are spin allowed by an exchange mechanism. The first of these processes is also allowed by coulombic interaction and at close distances both interactions could take place.

The second process is however forbidden by a coulombic mechanism and there are no compensating effects in this case since it is the $S_0 \to T_1$ transition in the acceptor which is forbidden. This process has become known as triplet–triplet energy transfer. It was first observed by Terenin and Ermolaev[52], who studied the quenching of phosphorescence from donor molecules and the sensitization of acceptor phosphorescence in rigid media. For all the donor and acceptor combinations used they found that the ratio of the phosphorescence yield in the absence and presence of acceptor concentration C_A followed the equation

$$\frac{\phi_D{}^0}{\phi_D} = e^{\alpha C_A} \tag{3.21}$$

Perrin first derived an equation of this form by assuming that the probability of quenching was unity within an 'active sphere' with the acceptor molecule at its centre. Donor molecules outside this sphere were assumed to be unaffected. Experimental values for R, the radius of the sphere, are given in Table 3.13. The values are not much greater than normal collisional diameters. The $S_0 \rightarrow T_1$ transition probabilities of the 1-halonaphthalenes used as acceptors vary by a thousand-fold without

Table 3.13 [a]

Volumes and radii of spheres of action for triplet–triplet energy transfer in a rigid alcohol/ether glass at 93° or 77°K

Donor	Acceptor	Volume of active sphere $(\times 10^{21}$ cm$^3)$	R (Å)
Benzaldehyde	Naphthalene	6·8	12
Benzaldehyde	1-Chloronaphthalene	7·0	12
Benzaldehyde	1-Bromonaphthalene	7·2	12
Benzophenone	Naphthalene	8·6	13
Benzophenone	1-Methylnaphthalene	9·5	13
Benzophenone	1-Chloronaphthalene	9·5	13
Benzophenone	1-Iodonaphthalene	8·6	13
Benzophenone	Quinoline	7·2	12
Acetophenone	Naphthalene	6·0	11
p-Chlorobenzaldehyde	Naphthalene	6·7	12
p-Chlorobenzaldehyde	1-Bromonaphthalene	6·2	11
o-Chlorobenzaldehyde	Naphthalene	5·4	11
m-Iodobenzaldehyde	Naphthalene	5·8	11
m-Iodobenzaldehyde	1-Bromonaphthalene	5·7	11
Xanthone	Naphthalene	9·2	13
Anthraquinone	Naphthalene	5·9	11
Anthraquinone	1-Bromonaphthalene	∼7·6	∼12
Triphenylamine	Naphthalene	9·3	13
Carbazole	Naphthalene	14	15
Phenanthrene	Naphthalene	10	13
Phenanthrene	1-Chloronaphthalene	11	14
Phenanthrene	1-Bromonaphthalene	∼11	∼14

[a] Values taken from Ermolaev[135].

affecting the transfer efficiency. These results are consistent with energy transfer by an exchange mechanism as are even more refined treatments of these data[53].

Table 3.14

Rate constants for triplet–triplet energy transfer as a function of the difference in triplet energy levels of the donor and acceptor, ΔE

Donor	Acceptor	Solvent	$\Delta E(\mathrm{cm}^{-1})$	$k_t(\mathrm{l \cdot mole^{-1} s^{-1})}$ $\times 10^{-8}$	Ref.
Biacetyl	3,4-Benzpyrene	Benzene	5000	82	136
Biacetyl	Anthracene	Benzene	5000	81	136
Biacetyl	1,2-Benzanthracene	Benzene	3200	70	136
Biacetyl	Pyrene	Benzene	3000	75	136
Triphenylene	Naphthalene	Hexane	2200	13	137
Biacetyl	trans-Stilbene	Benzene	2000	44	136
Phenanthrene	1-Iodonaphthalene	Hexane	1100	70	137
Phenanthrene	1-Bromonaphthalene	Hexane	900	1·5	137
Biacetyl	Coronene	Benzene	700	2·0	136
Biacetyl	1-Nitronaphthalene	Benzene	500	1·1	136
Phenanthrene	Naphthalene	Hexane	300	0·029	137
Biacetyl	2,2'-Binaphthyl	Benzene	150	0·097	136

Overlap of orbitals is required for exchange interactions so that in fluid media the maximum rate constants for electronic energy transfer occurring by this mechanism will be equal to those expected for diffusion controlled reactions. Some observed rates for triplet–triplet energy transfer are given in Table 3.14. When the triplet level of the donor is $1000 \mathrm{~cm}^{-1}$ or more above that of the acceptor the transfer rate constants approach the values calculated using the equation

$$k_D = \frac{8RT}{2000\eta} \mathrm{l \cdot mole^{-1} s^{-1}} \qquad (3.22)$$

where η is the solvent viscosity. Porter and Osborne[54] have shown that this equation gives good agreement with experiment for reactions which are diffusion controlled. There is a decrease in the measured rate constant for energy transfer as the triplet levels approach each other due to temperature dependent transfer in the reverse direction.

Triplet–triplet energy transfer has been much used to determine the photoreactive state in many photochemical reactions[55,56,57,58]. The principle of the method is as follows. Consider a photoreactive molecule, M. One can arrange for it to act as a triplet energy donor,

$$\text{(a)} \quad M^*(T_1) + A(S_0) \rightarrow M(S_0) + A^*(T_1)$$

or as an acceptor

$$\text{(b)} \quad D^*(T_1) + M(S_0) \rightarrow D(S_0) + M^*(T_1)$$

In each case partners should be chosen which allow the relative energy levels of each donor–acceptor pair to be as shown in Figure 3.14. A filter should be used to ensure that only the donor absorbs light. If process (a) occurs with high efficiency it should be possible to add enough acceptor so that the decay of the triplet state of M is all due to energy transfer. If the triplet acceptor molecules are chemically inert then any remaining reaction can be attributed to the singlet state of M, and any reactions which

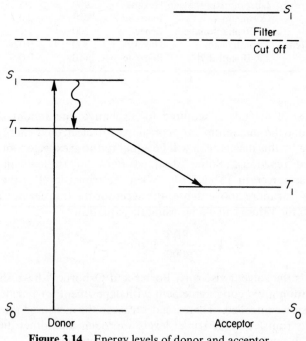

Figure 3.14   Energy levels of donor and acceptor

have been inhibited may be assigned to triplet state reactions. The presence, lifetime and heights of triplet states which decay too rapidly to be detected by flash photolysis or by other techniques have been established in this way.

Process (b) leads to photosensitized production of the triplet state of M bypassing excited singlet states. This allows a study to be made of the chemistry of the triplet state of M. By choosing donors which give high quantum yields of triplet state production, much higher yields of photoreaction of M can sometimes be obtained than those found when M itself is irradiated. The triplet state has been shown to be the photoreactive state in many reactions using this method[59,60]. Hammond and Lamola[10] have measured the yields of photosensitized reactions resulting from triplet–triplet energy transfer in order to calculate triplet state formation efficiencies. These are compared in Table 3.15 with values obtained by other methods.

It is extremely important with all these applications to confirm that triplet–triplet energy transfer is the only process occurring. Other processes are also capable of inhibiting and sensitizing reactions. Examples which have already proved troublesome include sensitization reactions of ketones which have been shown to be due to hydrogen atom transfer[61] rather than to triplet energy transfer[62]. Also, the value given in Table 3.15 for ϕ_T for naphthalene obtained from energy transfer measurements is now known to be too low because it has subsequently been discovered that the acceptor also quenches the singlet state of naphthalene by an as yet unexplained quenching process.

Dubois and coworkers[63,64,65] have shown that when long range interaction is not expected singlet–singlet transfer becomes a diffusion controlled process. Equation (3.22) predicts $k_D = 3 \cdot 5 \times 10^{10} \, \text{l} \cdot \text{mole}^{-1} \, \text{s}^{-1}$ for hexane at 28°C and this is very close to the values found (see Table 3.16).

The process

$$D^*(T_1) + A^*(T_1) \rightarrow D(S_0) + A^*(S_1)$$

where D and A may be the same or different species, is allowed by an exchange mechanism and is an example of what has become known as triplet–triplet annihilation which can give rise to delayed fluorescence from A^*. This delayed fluorescence depends on the square of the light intensity[66,67] and when donor and acceptor are the same species it has, at low triplet intensities, a lifetime equal to one-half the lifetime of the triplet state[68]. This process often takes place via an intermediate 'excimer' (see below) and delayed excimer fluorescence has also been observed[69] (see Figure 3.15). Triplet–triplet annihilation can also take place by a coulombic interaction[70] as mentioned earlier but in that case the products are

Table 3.15

Comparison of ϕ_T values obtained using Equation (3.24) and by other methods

Compound	Solvent	Values obtained using Equation (3.24)	Values obtained by other methods
Fluorene	Ethanol	0·32[a]	—
	Benzene	—	0·31[c]
Naphthalene	Ethanol	0·80[a]	0·71[b]
	Benzene	0·82[a]	0·40[c]
1-Methoxynaphthalene	Ethanol	0·50[a]	0·46[b]
	Benzene	—	0·26[c]
Acenaphthene	Ethanol	0·58[a]	0·45[b]
	Benzene	—	0·47[c]
Anthracene	Ethanol	0·72[a]	—
	Liquid paraffin	0·75[d]	0·58[e]
Phenanthrene	Ethanol	0·85[a]	0·80[b]
	Benzene	—	0·76[c]
	3-Methylpentane	—	0·70[e]
1,2-Benzanthracene	Ethanol	0·82[a]	—
	Hexane	0·77[a]	0·50[f]
Chrysene	Ethanol	0·85[a]	0·82[b]
	Benzene	—	0·67[c]
Pyrene	Ethanol	0·38[d]	0·27[b]
Coronene	Ethanol	0·56[a]	—

[a] Horrocks and Wilkinson[19].
[b] Parker and Joyce[72].
[c] Lamola and Hammond[10].
[d] Medinger and Wilkinson[11].
[e] Bowers and Porter[138].
[f] Labhart[139].

different. Delayed fluorescence in solution has been reviewed recently and some of the remaining problems have been discussed by Parker[71]. Parker and Joyce[72] have obtained values of ϕ_T from sensitized delayed fluorescence measurements (see Table 3.15).

Table 3.16

Rate constants for singlet–singlet energy transfer to biacetyl in aerated hexane solution at 28°C

Donor	Quenching constant[a] $K_Q(\text{l}\cdot\text{mole}^{-1})$	Mean lifetime[b] $10^9\tau_D(\text{s})$	$k_t(\text{l}\cdot\text{mole}^{-1}\text{s}^{-1}$ $\times 10^{-10})$
Benzene	214	5·7	3·8
Toluene	214	5·8	3·7
o-Xylene	208	6·0	3·5
m-Xylene	200	6·0	3·3
p-Xylene	210	6·1	3·4
Ethylbenzene	208	5·7	3·6
Cumene	204	6·0	3·4
Pentamethylbenzene	177	3·9	4·5
n-Propylbenzene	217	5·2	4·2
n-Butylbenzene	214	6·8	3·2
Hexamethylbenzene	80	2·0	4·0
Naphthalene	186	8·3	2·2

[a] Dubois and van Hemert[65].
[b] Ivanova, Kudryashov and Sveshnikov[140].

Exciton Migration

In molecular aggregates where there is often strong interaction between excited and unexcited molecules, the excitation energy may be regarded as being delocalized over large areas or alternatively the excitation energy may be thought of as 'hopping' from one molecule to another. A discussion of exciton migration would require many chapters. We will simply note here that in crystals migration of the excitation energy is extremely efficient with indications[73] of transfer rates as high as 10^{15} s^{-1}. In fluid media it is often difficult to distinguish excitation diffusion from material diffusion followed by collisional transfer or from single step long range transfer. However, a number of recent measurements suggest that with benzene as solvent, for example, solvent–solvent exciton transfer is detectable and accounts for measured transfer rates which are in excess of those expected as a result of material diffusion[74,75,76].

3.5.4 Quenching of Excited States

A process which increases the rate of decay of an electronically excited state is said to quench that state. Quenching may result from energy transfer (see Section 3.5.3), chemical reaction (see Section 3.5.5) or other bimolecular collisional processes, some of which are outlined below. The susceptibility of an excited state to quenching by species in the surrounding

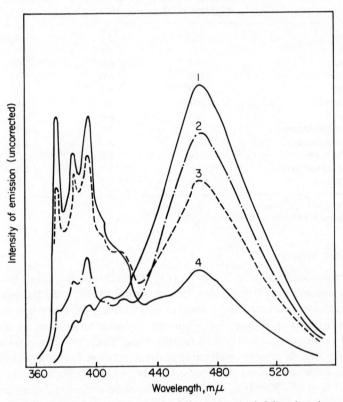

Figure 3.15 Delayed monomer fluorescence and delayed excimer fluorescence from pyrene in ethanol (1) $3 \times 10^{-3}\,M$, (2) $10^{-3}\,M$, (3) $3 \times 10^{-4}\,M$, (4) $2 \times 10^{-6}\,M$ (Parker and Hatchard[69])

medium is a function of its lifetime. A general relationship showing the effect of quenchers on molecular fluorescence intensity may be derived from a consideration of reactions 1 to 9.

Description	Reaction	Rate Constant
1. Excitation	$^1A + hv \to {}^1A^{*\prime}$	I_a einstein $l^{-1}\,s^{-1}$
2. Internal Conversion	$^1A^{*\prime} \to {}^1A^*$	$k_{S_X \to S_1}\,s^{-1}$
3. Fluorescence	$^1A^* \to {}^1A + hv_F$	$k_F\,s^{-1}$
4. Intersystem Crossing	$^1A^* \to {}^3A^*$	$k_{S_1 \to T}\,s^{-1}$
5. Internal Conversion	$^1A^* \to {}^1A$	$k_{S_1 \to S_0}\,s^{-1}$
6. Quenching of S_1	$^1A^* + Q \to$ quenching	$k_Q\,l \cdot mole^{-1}\,s^{-1}$
7. Phosphorescence	$^3A^* \to {}^1A + hv_p$	$k_P\,s^{-1}$
8. Intersystem Crossing	$^3A^* \to {}^1A$	$k_{T_1 \to S_0}\,s^{-1}$
9. Quenching of T_1	$^3A^* + Q \to$ quenching	$k'_Q\,l \cdot mole^{-1}\,s^{-1}$

where $^1A^{*\prime}$ represents any excited state, $^1A^*$ and $^3A^*$ the first excited singlet and triplet states respectively and I_a is the rate of light absorption. Under stationary state conditions

$$\frac{d[^1A^*]}{dt} = I_a - (k_F + k_{S_1 \to T} + k_{S_1 \to S_0} + k_Q[Q])[^1A^*] = 0$$

The quantum yield of fluorescence in the presence of the quencher is given by

$$\phi_F = \frac{k_F}{k_F + k_{S_1 \to T} + k_{S_1 \to S_0} + k_Q[Q]}$$

The ratio of the fluorescence yields in the absence and presence of quencher is equal to

$$\frac{\phi_F{}^0}{\phi_F} = 1 + \tau k_Q[Q]$$

where $\tau = 1/(k_F + k_{S_1 \to T} + k_{S_1 \to S_0})$ is the lifetime of $^1A^*$ in the absence of the quencher.

If F and F^0 are the fluorescence intensities at a convenient wavelength in the presence and absence of quencher, and provided the presence of the quencher causes no shift in the absorption or fluorescence spectra, $\phi_F{}^0/\phi_F = F^0/F$ and this gives the Stern–Volmer equation

$$\frac{F^0}{F} = 1 + K_{sv}[Q] \qquad (3.23)$$

where $K_{SV} = k_Q\tau = k_Q\tau_R\phi_F{}^0$ is the Stern–Volmer constant. This simple relationship is derived on the basis of direct excitation of ^1A (e.g. no

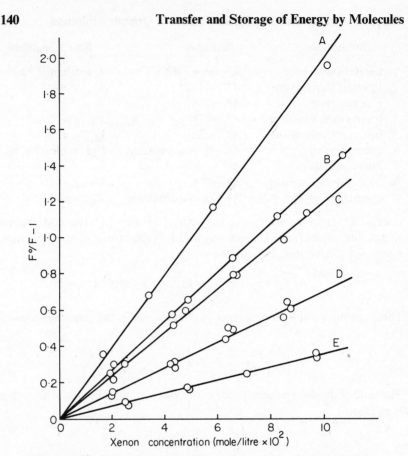

Figure 3.16 Quenching of the fluorescence of five aromatic hydrocarbons by xenon in 95 % ethanol: A = Chrysene, B = Phenanthrene, C = Coronene, D = Fluorene, E = Benzene (Horrocks, Kearvell, Tickle and Wilkinson[77])

energy transfer), a single emitting excited state and no absorption by the quencher. It gives good agreement with experiment in many cases (see for example Figure 3.16).

Heavy Atom Quenching

When fluorescence is quenched by compounds containing heavy atoms it can be shown that there is a corresponding increase in the amount of triplet state produced by observing the intensity of triplet–triplet absorption following flash excitation[11].

If reaction 5 is replaced by reaction 10

10. $^1A^* + Q \rightarrow \, ^3A^* + Q$

then in the absence of the quencher, the quantum yield of the triplet state formation is given by

$$\phi_T^0 = \frac{k_{S_1 \rightarrow T}}{k_F + k_{S_1 \rightarrow S_0} + k_{S_1 \rightarrow T}}$$

and in the presence of quencher $[Q]$

$$\phi_T = \frac{k_{S_1 \rightarrow T} + k_Q[Q]}{k_F + k_{S_1 \rightarrow S_0} + k_{S_1 \rightarrow T} + k_Q[Q]}$$

It follows therefore that

$$\frac{\phi_T}{\phi_T^0} = \frac{[^3A^*]}{[^3A^{0*}]} = \frac{1 + k_Q[Q]/k_{S_1 \rightarrow T}}{1 + k_Q \tau[Q]} \tag{3.24}$$

where $[^3A^{0*}]$ and $[^3A^*]$ are the concentrations of the triplet state produced initially following the exciting flash and in the absence and presence of quencher respectively. If the extinction coefficient of the triplet–triplet absorption maximum is independent of quencher concentration then

$$\frac{[^3A^*]}{[^3A^{0*}]} = \frac{D_T}{D_T^0}$$

where D_T and D_T^0 are the initial optical densities at the absorption maxima with and without added quencher.

By introducing the Stern–Volmer equation into Equation (3.24) an expression can be obtained which contains no term for the quencher concentration,

$$\frac{F^0}{F} - 1 = \phi_T^0 [(D_T F^0 / D_T^0 F) - 1] \tag{3.25}$$

Plots of F^0/F against $(D_T F^0 / D_T^0 F) - 1$ should give straight lines of slope ϕ_T^0 independent of the quencher used. Figure 3.17 shows the results obtained for 9-phenylanthracene in ethylene glycol. Xenon has been used as a heavy atom quencher in ethanol solution in which it dissolves to the extent of 0·1 mole litre^{-1} for a partial pressure of about one atmosphere.

Xenon is an ideal quencher[77] as it does not itself absorb, chemically react or have any effect on the absorption or emission spectra of the solute (see Figures 3.16 and 3.18).

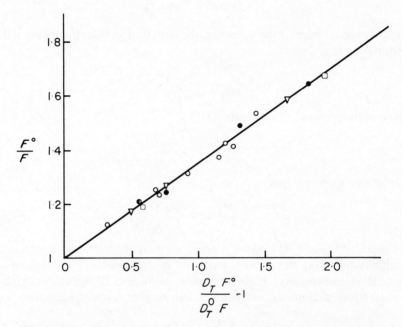

Figure 3.17 Plot showing relationship between relative fluorescence intensities and relative initial amounts of triplet-state production for 9-phenylanthracene in liquid paraffin with various amounts of quencher, ○, bromobenzene, ●, bromocyclohexane, ◑, p-dibromobenzene, ▫, ethyle iodide, and ▽, n-propyl iodide (Horrocks, Medinger and Wilkinson[142])

Heavy atom containing compounds also quench triplet states in solution (reaction 9) but fortunately the quenching rate constants are much lower than for excited singlet states. It is this disparity which makes possible the measurement of the enhanced triplet–triplet absorption by flash photolysis. As has been mentioned in Section 3.4, heavy atoms tend to decrease the measured phosphorescence lifetime of the triplet state $\tau_{T,}$ and increase the phosphorescence yield ϕ_P. Since diffusional quenching is negligible in the rigid media in which most phosphorescence studies are carried out, reactions 7 and 8 determine the measured lifetime

$$\tau_{T_1} = \frac{1}{k_P + k_{T_1 \to S_0}}$$

Using combined ESR and optical techniques, Siegel and Judeikis[78] have shown that in rigid media the phosphorescence process is more sensitive to external heavy atoms than is intersystem crossing from T_1.

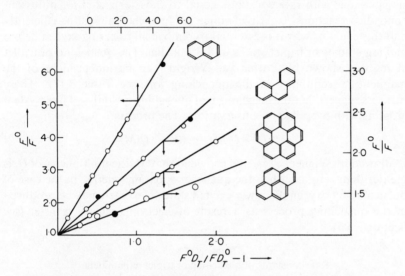

Figure 3.18 Plots for the determination of ϕ_T of four aromatic hydrocarbons using xenon as a heavy atom quencher.

Naphthalene	○	6.2×10^{-5} M	●	4.9×10^{-5} M
Phenanthrene	○	5×10^{-5} M	●	3×10^{-5} M
Coronene	○	5.9×10^{-6} M		
Pyrene	○	2×10^{-5} M	●	1.45×10^{-5} M

The increase in intersystem crossing from $S_1 \to T_1$ due to the presence of heavy atoms should be useful in determining the reactive state in many photochemical reactions in solution, since the yields of reactions occurring via the triplet state should increase and reaction involving excited singlet states should decrease in the presence of heavy atom quenchers. Xenon would make an ideal quencher for this purpose. Recently the reactive state in the photoreduction of acridine has been shown to be the first excited singlet state since the quantum yield of the reaction of acridine in ethanol decreases in the presence of iodide ions by the same amount as the fluorescence[79].

Quenching by Paramagnetic Species

Paramagnetic molecules and ions reversibly quench the excited singlet and triplet states of many compounds. Oxygen and nitric oxide quench singlet states with rate constants equal to those expected for diffusion-controlled reactions[80]. Triplet states are quenched almost as efficiently and this explains why it is essential to remove all traces of oxygen before making studies of triplet states in fluid media. The quenching of triplet states was shown by Porter and Wright[81] to be independent of the magnetic susceptibility of the quenching ion (see Table 3.17). These workers suggest that the mechanism of quenching is catalysed intersystem crossing with overall spin conservation. The process

$$A^*(T_1) + Q(M) \rightarrow A(S_0) + Q(M)$$

is allowed by Wigner's spin rule provided that M, the multiplicity of Q, is greater than zero, i.e. when the quencher is paramagnetic. In the case of quenching by oxygen the lower excited states of oxygen may be produced in the quenching process as a result of electronic energy transfer (see Section 3.5.5).

Table 3.17 [a]

Rate constants of quenching of triplet naphthalene
by ions in water and ethylene glycol

Ion	$k_Q(\text{l} \cdot \text{mole}^{-1} \text{s}^{-1} \times 10^{-7})$ in water	in ethylene glycol	No. of unpaired electrons	Paramagnetic susceptibility (Bohr magnetons)
Cu^{2+}	7·5	7·3	1	1·93
Ni^{2+}	2·3	2·4	2	3·21
Co^{3+}	5·0	4·4	3	5·01
Cr^{3+}	6·9	—	3	3·82
Fe^{2+}	—	3·8	4	5·30
Fe^{3+}	2·9	—	5	5·85
Mn^{2+}	2·8	1·6	5	5·81
Nd^{3+}	—	0·04	3	3·60
Gd^{3+}	—	0·007	7	8·01

[a] Data from Porter and Wright[81].

Excimers and Exciplexes

The quantum yields of fluorescence of most fluorescent substances in solution decrease with increasing concentration. In many cases the quenching is due to collisions between ground state and excited singlet state molecules which form an excited dimeric state (excimer) which may dissociate, with or without emission of fluorescence, or as in anthracene, pass over to give a dimer which is stable in the ground state (see Section 3.5.5.). Excimer emission has been observed as a broad, structureless band to the red of the monomer fluorescence for a large number of aromatic hydrocarbons[82,83,84,85], e.g. pyrene (see Figure 3.19), 1,2-benzanthracene, 1-methylanthracene and naphthalene. The appearance of excimer fluorescence is not accompanied by a change in the absorption spectrum showing that excimers are only stable in the excited state. It has been suggested that excimer formation is a special case of molecular complex formation in the excited state. Excited complexes or 'exciplexes' can also be formed between pairs of molecules which may have quite diverse chemical structures, e.g. biphenyl and diethylaniline[86]. Aromatic exciplex formation has been shown to be due to charge-transfer interaction. In a flash photolysis study of the quenching of perylene fluorescence by amines, Leonhardt and Weller[87] found a very pronounced transient absorption at 580 mμ and identified it with the perylene monoanion. They suggested that in polar solvents the excited charge transfer complex could dissociate into ions. In organic solvents emission from the exciplex competes with monomer fluorescence,

$$A^*(S_1) + B(S_0) \rightleftharpoons (A^- B^+)^*(S_1) \xrightarrow[\text{solvent}]{\text{polar}} A^- + B^+$$

$$\downarrow h\nu_m \qquad\qquad \downarrow h\nu_e$$

$$A(S_0) \qquad\qquad A(S_0) + B$$

becoming more effective as the concentration of quencher is increased (see Figure 3.20).

3.5.5 Photochemical Reactions

Because molecular excited states have physical properties different from the ground state, including a considerable amount of additional energy, it is to be expected that they will undergo chemical reactions which do not occur or which occur to different extents in the ground state. However the rate of reaction of an excited state must be sufficiently fast to compete

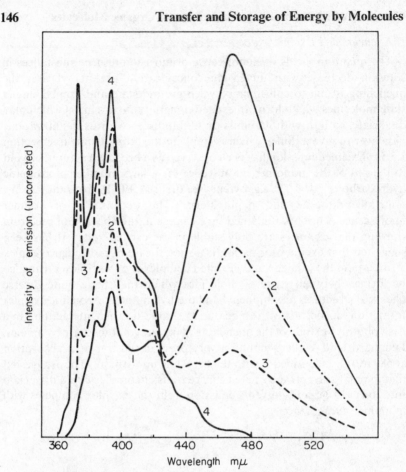

Figure 3.19   Normal fluorescence from pyrene monomer and excimer in ethanol: (1) $3 \times 10^{-3}\,M$, (2) $10^{-3}\,M$, (3) $3 \times 10^{-4}\,M$, (4) $2 \times 10^{-6}\,M$ (Parker and Hatchard[69])

effectively with deactivation by photophysical processes. From the comments made regarding radiative and radiationless transition probabilities in Section 3.4 it follows that, except for very rapid unimolecular reactions such as dissociations and some isomerizations, almost all reactions will be due to either the first excited singlet or triplet states. Since the lifetime of the lowest triplet state in solution is usually about 10^5 times longer than that of the lowest excited singlet state it is often suggested that the majority

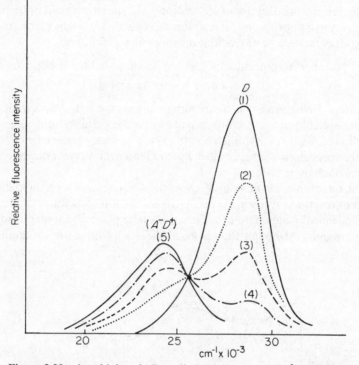

Figure 3.20   A = biphenyl; D = diethylaniline. (1) $10^{-2}\ M$ diethylaniline (in toluene), (2) $10^{-2}\ M$ diethylaniline + 0·03 M biphenyl; (3) $10^{-2}\ M$ diethylaniline + 0·10 M biphenyl, (4) $10^{-2}\ M$ diethylaniline + 0·30 M biphenyl; (5) $10^{-2}\ M$ diethylaniline + high concentration biphenyl (Knibbe and Weller[143])

of photochemical reactions involve triplet states. While this may generally be true, reactions involving excited singlet states can and do occur and the greater energy available compared with the triplet state may accelerate their reactions. Therefore it is necessary to characterize the photoreactive state before a reaction mechanism can be postulated.

Most photochemical reactions are complex and occur in many stages thus only in rare cases is the measured change that which was originally produced by the absorption of light. Einstein's law of chemical equivalents, which states that one quantum of radiation is absorbed by each molecule taking part in the chemical reaction induced by light, can therefore only be applied to the primary act of absorption. The quantum yield or quantum efficiency of a photochemical reaction ϕ_R is defined in terms of the amount of product formed or of reactant disappearing as follows

$$\phi_R = \frac{\text{Number of molecules of reactant decomposed (or product formed)}}{\text{Number of quanta absorbed}}$$

Quantum yield values which are simple integers, e.g. 1, 2 or 3 indicate a simple stoichiometric relationship between the primary and secondary reactions. A value very much less than one indicates some efficient competitive processes whereas values much greater than one indicate photochemical chain reactions.

The quantum yield of reaction should not be confused with the chemical yield of reaction. A low quantum yield of reaction sometimes can lead to high chemical yields if no other reactions take place and if one irradiates long enough. Although the chemical yield is of little use in helping to

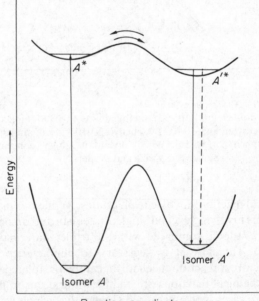

Figure 3.21 Schematic diagram illustrating photochemical route for isomerization, $A \rightarrow A'$

establish the mechanism of a photochemical reaction in photopreparative work a high chemical yield may be of the utmost importance.

There have been a number of excellent books[88,89,90,91] and reviews[92,93,94,95] which have appeared recently on the subject of organic photochemical reactions and space does not allow a comprehensive list here. However it is hoped that the following examples will illustrate something of the variety of types of photochemical reactions occurring in solution and the mechanisms involved.

Cis–Trans Isomerization

Irradiation of solutions of either geometrical isomer of an olefin with ultraviolet or visible light often results in a certain amount of *cis–trans* isomerization[96,97]. In contrast with thermal and catalytic isomerization, the photochemical reaction often favours formation of the less thermodynamically stable isomer which is usually the *cis* isomer. Figure 3.21 shows schematically how this is possible. Continued irradiation leads to the formation of a photostationary mixture in which the ratio of the two isomers remains constant.

Some of the most extensively studied photoisomerizations are those of stilbene and its derivatives but the mechanisms of these reactions are still imperfectly understood[98,99,100,101,102].

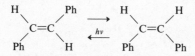

It was originally thought that interconversion between the isomers occurred in an excited state common to both and produced directly by absorption. However the fact that *trans*-stilbene shows appreciable fluorescence, whereas the fluorescence from *cis*-stilbene is very weak, eliminates this possibility[98]. The quantum yield of isomerization is independent of the wavelength of the exciting light for both species indicating that higher excited singlet states are not involved in the reaction[100].

These considerations led to the suggestion of a common lowest triplet state, T_1, in which rotation about the carbon–carbon axis could occur. Quantum mechanical calculations on ethylene showed that there are minima in the curves of potential energy versus the angle of twist about the carbon–carbon bond for both S_1 and T_1 when the planes of the methylene groups are perpendicular to each other[103]. However for complex olefins such as stilbene where the double bond is part of a conjugated π system the system is more complicated and there does not seem to be an electronic state common to both isomers.

From singlet–triplet absorption spectra in a heavy atom solvent, Dyck and McClure[98] have found the energies of the lowest triplet states to be 50 kcal mole^{-1} for *trans*-stilbene and 57 kcal mole^{-1} for *cis*-stilbene. When the triplet states are populated by energy transfer the *cis/trans* ratio depends on the transfer efficiency to the two T_1 states, e.g. if the triplet energy of the donor is >62 kcal mole^{-1}, transfer to both triplets is diffusion-controlled and the *cis/trans* ratio remains constant at 0·45. Hammond and Saltiel[104] have postulated the existence of a 'phantom triplet' almost isoenergetic with the *trans* triplet but whose geometry differs considerably from either *cis* or *trans* spectroscopic triplet states. They suggest the phantom triplet may not be populated by direct absorption because it would contravene the Franck–Condon principle but it may be populated by triplet–triplet energy transfer from sensitizers with triplet state energies lying between 40 and 57 kcal mole^{-1}.

Photochemical Cycloadditions and Dimerizations

Illumination of a deoxygenated solution of benzoquinone in 2,3-dimethylbuta-1,3-diene with blue light gives a spiro-pyran in 33% yield[105].

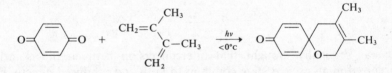

Apparently this type of reaction occurs only if triplet–triplet energy transfer from the carbonyl compound (in this case quinone) to the diene cannot take place, i.e. if the diene has the higher triplet state energy.

Dimerization of stilbenes to form tetraphenylcyclobutane does not occur at all when excitation is accomplished by triplet–triplet energy transfer. However if the pure compound in a very concentrated solution is irradiated with ultraviolet light a small amount of dimer is formed.

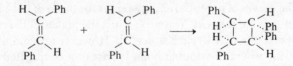

Saltiel[92] suggests that the first excited singlet state is responsible for this dimerization.

Anthracene molecules in solution begin to dimerize above certain concentrations with a concomitant reduction of fluorescence emission. The quantum yield of dianthracene formation in degassed benzene reaches a maximum of 0·3 at which concentration the fluorescence yield is effectively zero[106]. This suggests that the formation of dianthracene involves a ground state and a first excited singlet state molecule, an idea which is supported by the negligible dianthracene formation in bromobenzene solution where the spin–orbit coupling effect of the heavy atom markedly quenches fluorescence.

Bäckström and Sandros[107] have shown that some dianthracene formation occurs when anthracene is excited by triplet–triplet energy transfer from biacetyl in degassed benzene solution. The biacetyl triplet energy is considerably greater than that of anthracene and energy transfer is diffusion controlled. Nevertheless the rate of sensitized dimerization is extremely low indicating that dianthracene formation occurs only following triplet–triplet annihilation, i.e.

$$A^*(T_1) + A^*(T_1) \rightarrow {}^1A^*(S_1) + A(S_0) \rightarrow A_2$$

The anthracene units of the dianthracene molecule are joined across their meso positions with a mutual loss of planarity due to the change in carbon hybridization. Dianthracene-type dimers are also formed by irradiation of 9-substituted anthracenes and it appears that some dimers favour the 'syn' or head-to-head arrangement (IV) e.g. with R = CHO, CH_2OH or CO_2CH_3,[108] while dipole moment studies show that others favour the 'anti' or head-to-tail arrangement (V), e.g. with R = Br or Cl.[109]

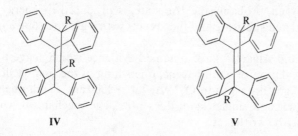

IV   V

When both meso positions are substituted in anthracene no stable photodimers of this type are formed. This is attributed to the steric effects of the substituents. However a stable mixed photodimer of anthracene and 9,10-dichloroanthracene has been obtained[110].

If the dianthracene-type molecules are dispersed in a rigid matrix and irradiated with light of wavelength 2537 Å at liquid nitrogen temperatures the meso–meso bonds are broken. However the anthracene molecules remain paired in a 'sandwich' arrangement which gives rise to excimer emission on subsequent excitation[111]. The structure is irreversibly transformed to ordinary anthracene when the solution is warmed.

Photoperoxidation

As well as quenching excited states by a photophysical process oxygen sometimes photochemically reacts with the substrate as in the well-known photoperoxidation of anthracene which yields the endoperoxide (VI)

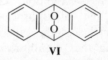

VI

The yield of photoperoxidation increases with increasing anthracene concentration. The participation of an excited state of oxygen in these reactions was originally suggested by Kautsky[112]. More recently Foote and Wexler[113] have shown that the peroxides formed by the reaction of acenes with chemically produced singlet states of oxygen are identical with the photoperoxides. Wilson[114] has obtained kinetic evidence which indicates a common excited species in these reactions. Corey and Taylor[115] have bubbled oxygen which has been subjected to an electrodeless discharge through solutions of anthracene and some of its derivatives and have produced the endoperoxides. All this suggests that an excited state of oxygen, probably the $^1\Delta g$ state, is produced by energy transfer from the excited substrate and then reacts with a ground state anthracene molecule.

Stevens and Algar[116] have obtained evidence which suggests that, in photoperoxidation of naphthacene, quenching of the first excited triplet state by oxygen does lead to O_2^* $^1\Delta g$ (at $\approx 8000\ cm^{-1}$) but quenching of the lowest excited singlet state does not. The mechanism which they support may be written as

1. $$A(S_0) + h\nu \rightarrow A^*(S_1)$$

2. $$A^*(S_1) \rightarrow A^*(T_1)$$

3. $$A^*(S_1) \rightarrow A(S_0) + h\nu_F$$

4. $$A^*(S_1) \rightarrow A(S_0)$$

5. $$A^*(S_1) + O_2(^3\Sigma g) \rightarrow A^*(T_1) + O_2(^3\Sigma g)$$

6. $$A^*(T_1) \rightarrow A(S_0)$$

7. $$A^*(T_1) + O_2(^3\Sigma g) \rightarrow A(S_0) + O_2^*(^1\Delta g)$$

8. $$A(S_0) + O_2^*(^1\Delta g) \rightarrow AO_2$$

9. $$O_2^*(^1\Delta g) \rightarrow O_2(^3\Sigma_g)$$

The process

$$A^*(S_1) + O_2(^3\Sigma g) \rightarrow A^*(T_1) + O_2^*(^1\Delta g)$$

is energetically possible and obeys Wigner's spin rule and yet catalysed intersystem crossing step 5 is apparently preferred.

Photoreduction of Anthraquinone

When anthraquinone is irradiated in deaerated alcoholic solution it quantitatively converts to anthrahydroquinone[117] (see Figure 3.22). In the presence of oxygen, anthraquinone and several of its derivatives sensitize the oxidation of alcohols to their corresponding aldehydes or ketones[118,119,120]. In the presence of anthracene and naphthalene, which act as inert acceptors of triplet state energy, these reactions are inhibited and the sensitized production of the triplet states of the acceptors can be confirmed in flash photolysis experiments. This indicates that the lowest triplet state of anthraquinone, which is a $^3(n, \pi^*)$ state, is the photoreactive species. The following mechanism may be written for the reactions in the absence of oxygen.

1. $$Q(S_0) + h\nu \overset{I_A}{\rightarrow} Q^*(S_X)$$

2. $$Q^*(S_X) \rightarrow Q^*(S_1)$$

3. $$Q^*(S_1) \rightarrow Q^*(T_1)$$

4. $$Q^*(T_1) + R_2CHOH \rightarrow QH\cdot + R_2\dot{C}OH$$

5. $$R_2\dot{C}OH + Q(S_0) \rightarrow QH\cdot + R_2CO$$

6. $$2R_2\dot{C}OH \rightarrow R_2CO + R_2CHOH$$

7. $$2QH\cdot \rightarrow Q + QH_2$$

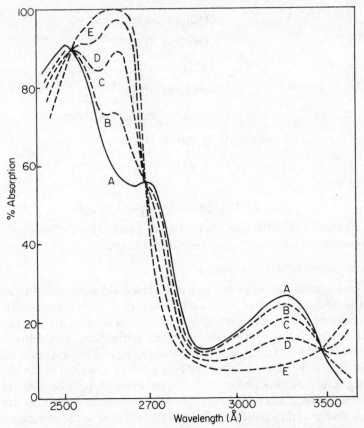

Figure 3.22 Variation in absorption due to irradiation of a $2.1 \times 10^{-5}\,M$ anthraquinone solution with 2537 Å excitation for (A) 0 s, (B) 25 s, (C) 75 s, (D) 180 s, (E) 480 s (Tickle and Wilkinson[117])

A similar mechanism with step 7 replaced by a dimerization reaction has been established for the photoreduction of benzophenone in alcoholic solution[121,122]. Assuming photostationary state conditions it can be shown that

$$\frac{1-\phi}{(2\phi-1)^2} = \frac{k_5 I_A}{(k_4 [Q])^2} \tag{3.26}$$

where ϕ is the quantum yield of production of anthrahydroquinone[117].

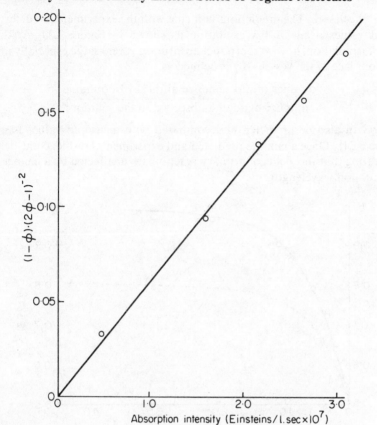

Figure 3.23   Plot of $(1-\phi)/(2\phi-1)^2$ against I_A, the rate of absorption of 3660 Å irradiation for an $8\cdot6 \times 10^{-5}$ M anthraquinone solution

With excitation at 3660 Å it is possible to make measurements at such low optical densities that absorption may be considered to be uniform throughout the reaction vessel. The intensity dependence of the quantum yields can then be plotted according to Equation (3.26) and this gives $k_5/k_4^2 = 4\cdot3 \times 10^{-3}$ s mole/l and confirms the linear relationship expected from the mechanism (see Figure 3.23). The concentration dependence can be predicted from a knowledge of the value of k_5/k_4^2 taking into account the variation of the rate of absorption along the path of the exciting light within the vessel. The predictions coincide with the experimentally determined values using 3130 Å excitation as shown in Figure 3.24. With 2537 Å irradiation, however, product absorption is very large especially at the front face of the vessel. If γ is defined as

$$\gamma = \frac{\text{number of molecules of anthrahydroquinone}}{\text{number of quanta absorbed in the solution}}$$

then γ can also be predicted and compared with measured values (see Figure 3.24). Once again the predicted and experimental results coincide, illustrating that the yields of primary reaction are unaffected by a change in excitation wavelength.

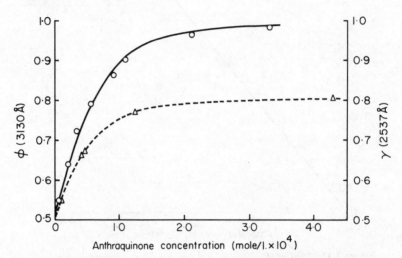

Figure 3.24 The dependence of ϕ and γ on anthraquinone concentration: ○ Experimental values; —, predicted curve for 3130 Å excitation; △ Experimental values for 2537 Å excitation (Tickle and Wilkinson[117])

In the presence of oxygen the following reactions account for the photosensitization of the alcohol and the lack of consumption of the quinone

8. $QH \cdot + O_2 \rightarrow \cdot HO_2 + Q$

9. $R_2\dot{C}OH + O_2 \rightarrow \cdot HO_2 + R_2CO$

10. $QH_2 + O_2 \rightarrow H_2O_2 + Q$

11. $\cdot HO_2 + HO_2 \rightarrow H_2O_2 + O_2$

The concentration of oxygen in a saturated solution is too small for the reaction

12. $Q^*(T) + O_2 \rightarrow$ quenching

to compete with reaction 4 and thus the very efficient photosensitized oxidation of alcohol by quinone is due to hydrogen atom transfer to oxygen and not to a process which involves electronic energy transfer, in contrast to the reaction discussed in the previous section.

REFERENCES

1. W. D. McElroy and H. H. Seliger, *Arch. Biochem. Biophys.*, **88**, 136 (1960).
2. *Pulse Radiolysis* (Ed. Ebert, Keene, Swallow and Baxendale), Academic Press, 1965.
3. H. H. Jaffe and M. Orchin, *Theory and Application of Ultraviolet Spectroscopy*, Wiley, New York, 1962.
4. W. Kauzmann, *Quantum Chemistry*, Academic Press, New York, 1957.
5. S. P. McGlynn, F. J. Smith and G. Cilento, *Photochem. Photobiol.*, **3**, 269 (1964).
6. R. Pariser, *J. Chem. Phys.*, **24**, 250 (1956).
7. R. Astier and Y. H. Meyer, *The Triplet State*, Cambridge University Press, 1967, p. 447.
8. D. R. Kearns and W. A. Case, *J. Am. Chem. Soc.*, **88**, 5087 (1966).
9. D. F. Evans, *J. Chem. Soc.*, **1959**, 2753.
10. A. A. Lamola and G. S. Hammond, *J. Chem. Phys.*, **43**, 2129 (1965).
11. T. Medinger and F. Wilkinson, *Trans. Faraday Soc.*, **61**, 620 (1965).
12. S. J. Strickler and R. D. Berg, *J. Chem. Phys.*, **37**, 814 (1962).
13. J. B. Birks and D. J. Dyson, *Proc. Roy. Soc.*, **A275**, 135 (1963).
14. C. A. Parker and C. G. Hatchard, *Trans. Faraday Soc.*, **57**, 1894 (1961).
15. V. L. Ermolaev, *Soviet Phys.—Usp.* (*English Transl.*), **6**, 333 (1963).
16. G. W. Robinson and R. P. Frosch, *J. Chem. Phys.*, **38**, 1187 (1963).
17. M. R. Wright, R. P. Frosch and G. W. Robinson, *J. Chem. Phys.*, **33**, 934 (1960).
18. F. Wilkinson and J. T. Dubois, *J. Chem. Phys.*, **39**, 377 (1963).
19. A. R. Horrocks and F. Wilkinson, *Proc. Roy. Soc.*, **A306**, 257 (1968).
20. J. Czekalla, *Z. Electrochem.*, **64**, 1221 (1960).
21. E. Lippert, *Z. Electrochem.*, **61**, 962 (1957).
22. N. Mataga, Y. Kaifa and M. Koizumi, *Bull. Chem. Soc., Japan.* **28**, 690 (1955), and **29**, 465 (1956).
23. E. G. McRae, *J. Phys. Chem.*, **61**, 562 (1959).
24. N. G. Bakhshiev, *Opt. Spectry.* (*English Transl.*), **10**, 379 (1961).
25. N. S. Bayliss and E. G. McRae, *J. Phys. Chem.*, **58**, 1002 (1954).
26. L. Onsager, *J. Am. Chem. Soc.*, **58**, 1486 (1936).
27. M. B. Ledger and P. Suppan, *Spectrochim. Acta*, **23**, 641 (1967).
28. D. H. Phelps and F. W. Dalby, *Can. J. Phys.*, **43**, 144, 1766 (1965).

29. D. E. Freeman and W. Klemperer, *J. Chem. Phys.*, **40**, 604 (1964) and **45**, 52 (1966).
30. D. E. Freeman, J. R. Lombardi and W. Klemperer, *J. Chem. Phys.*, **45**, 58 (1966).
31. J. R. Lombardi, D. Campbell and W. Klemperer, *J. Chem. Phys.*, **46**, 3482 (1967).
32. K. Weber, *Z. Phys. Chem.*, **B15**, 18 (1931).
33. Th. Förster, *Z. Electrochem.*, **54**, 42 (1950).
34. A. Weller, *Progress in Reaction Kinetics*, **1**, 187 (1961).
35. A. Weller, *Z. Electrochem.*, **56**, 662 (1952).
36. E. L. Wehry and L. B. Rogers, *Spectrochim. Acta*, **21**, 1976 (1965).
37. C. Sandorfy, *Compt. Rend.*, **232**, 841 (1951), and *Can. J. Chem.*, **31**, 439 (1953).
38. C. A. Coulson and J. Jacob, *J. Chem. Soc.*, **1949**, 1984.
39. D. W. Ellis and L. B. Rogers, *Spectrochim. Acta*, **18**, 265 (1962), and **20**, 1709 (1964).
40. J. C. Havlock, S. F. Mason and B. E. Smith, *J. Chem. Soc.*, **1963**, 4897.
41. A. Weller, *Z. Electrochem.*, **60**, 1144 (1956).
42. A. Weller, *Naturwiss.*, **7**, 175 (1955).
43. G. Jackson and G. Porter, *Proc. Roy. Soc.*, **A260**, 13 (1961).
44. Th. Förster, *Discussions Faraday Soc.*, **27**, 7 (1959).
45. S. A. Latt, H. T. Cheung and E. R. Blout, *J. Am. Chem. Soc.*, **87**, 995 (1965).
46. K. H. Drexhage, M. M. Zwick and H. Kuhn, *Ber. Bundes. Phys. Chem.*, **67**, 62 (1963).
47. C. Bojarski, *Ann. Physik.*, **12**, 253 (1963).
48. V. L. Ermolaev and E. B. Sveshnikova, *Dokl. Akad. Nauk. SSSR.* **149**, 1295 (1963).
49. R. G. Bennett, R. P. Schwenker and R. E. Kellogg, *J. Chem. Phys.*, **41**, 3040 (1964).
50. D. L. Dexter, *J. Chem. Phys.*, **21**, 836 (1953).
51. E. Wigner, *Nachr. Ges. Wiss. Göttingen, Math.-physik. Kl.*, **1927**, 375.
52. A. Terenin and V. L. Ermolaev, *Trans. Faraday Soc.*, **52**, 1042 (1956).
53. M. Inokuti and F. Hiryama, *J. Chem. Phys.*, **43**, 1978 (1965).
54. A. D. Osborne and G. Porter, *Proc. Roy. Soc.*, **A284**, 9 (1965).
55. G. Porter and F. Wilkinson, *Trans. Faraday Soc.*, **57**, 1686 (1961).
56. G. S. Hammond, N. J. Turro and P. A. Leersmakers, *J. Phys. Chem.*, **66**, 1144 (1962).
57. R. B. Cundall and A. S. Davies, *Trans. Faraday Soc.*, **62**, 2444 (1966).
58. G. S. Hammond and N. J. Turro, *Science*, **142**, 1541 (1963).
59. F. Wilkinson, *J. Phys. Chem.*, **66**, 2569 (1962).
60. G. S. Hammond, C. A. Stout and A. A. Lamola, *J. Am. Chem. Soc.*, **86**, 3103 (1964).
61. E. Vander-Donckt and G. Porter, *J. Chem. Phys.*, **46**, 1173 (1967).
62. A. Kellmann and J. T. Dubois, *J. Chem. Phys.*, **42**, 2518 (1965).
63. J. T. Dubois and B. Stevens, *Luminescence of Organic and Inorganic Materials*, Wiley, New York, 1962, p. 115.
64. J. T. Dubois and M. Cox, *J. Chem. Phys.*, **38**, 2536 (1963).
65. J. T. Dubois and R. L. Van Hemert, *J. Chem. Phys.*, **40**, 923 (1964).

66. C. A. Parker and C. G. Hatchard, *Proc. Roy. Soc.*, **A269**, 574 (1962).
67. C. A. Parker and C. G. Hatchard, *Proc. Chem. Soc.*, **1962**, 147.
68. B. Stevens and M. S. Walker, *Proc. Roy. Soc.*, **A281**, 420 (1964).
69. C. A. Parker and C. G. Hatchard, *Trans. Faraday Soc.*, **59**, 284 (1963).
70. R. E. Kellogg, *J. Chem. Phys.*, **41**, 3046 (1964).
71. C. A. Parker, *The Triplet State*, Cambridge University Press, 1967, p. 353.
72. C. A. Parker and T. A. Joyce, *Trans. Faraday Soc.*, **62**, 2785 (1966).
73. A. S. Davydov, *The Theory of Molecular Excitons*, McGraw-Hill, New York, 1962.
74. C. Lloyd Braga, M. D. Lumb and J. B. Birks, *Trans. Faraday Soc.*, **62**, 1830 (1966).
75. J. T. Dubois and J. W. Van Loben Sels, *The Physics and Chemistry of Scintillators*, Karl Thiemig Kg., Munich, 1966, p. 74.
76. M. A. Dillon and M. Burton in *Pulse Radiolysis* (Ed. Ebert, Keene, Swallow and Baxendale), Academic Press, 1965, p. 259.
77. A. R. Horrocks, A. Kearvell, K. Tickle and F. Wilkinson, *Trans. Faraday Soc.*, **62**, 3393 (1966).
78. S. Siegel and H. S. Judeikis, *J. Chem. Phys.*, **42**, 3060 (1965).
79. F. Wilkinson and J. T. Dubois, *J. Chem. Phys.*, **48**, 2651 (1968).
80. W. Ware, *J. Phys. Chem.*, **66**, 455 (1962).
81. G. Porter and M. R. Wright, *Discussions Faraday Soc.*, **27**, 18 (1957).
82. Th. Förster and K. Kasper, *Z. Electrochem.*, **59**, 977 (1955).
83. B. Stevens and E. Hutton, *Nature*, **186**, 1045 (1960).
84. J. B. Birks and L. G. Christophorou, *Proc. Roy. Soc.*, **A277**, 571 (1964).
85. J. B. Birks and J. B. Aladekomo, *Photochem. Photobiol.*, **2**, 415 (1963).
86. H. Beens, H. Knibbe and A. Weller, *J. Chem. Phys.*, **47**, 1183 (1967).
87. H. Leonhardt and A. Weller, *Luminescence of Organic and Inorganic Materials*, Wiley, 1962, p. 74.
88. N. J. Turro, *Molecular Photochemistry*, Benjamin, New York, 1965.
89. J. G. Calvert and J. N. Pitts, Jr., *Photochemistry*, Wiley, New York, 1966.
90. R. O. Kan, *Organic Photochemistry*, McGraw-Hill, New York, 1966.
91. *Advances in Photochemistry*, Vol. 1–4, (Ed. W. A. Noyes, Jr., G. S. Hammond and J. N. Pitts, Jr.), Interscience, 1963–1967.
92. J. Saltiel, *Survey of Progress in Chemistry*, **2**, 239 (1964).
93. 'International symposium in organic photochemistry' (Strasbourg 1964), *Pure Appl. Chem.*, **9**, 460–621 (1965).
94. J. M. Bruce, *Quart. Rev. (London)*, **21**, 405 (1967).
95. W. L. Dilling, *Chem. Rev.*, **66**, 373 (1966).
96. G. M. Wyman, *Chem. Rev.*, **55**, 625 (1955).
97. R. B. Cundall, *Progress in Reaction Kinetics*, **2**, 165 (1964).
98. R. H. Dyck and D. S. McClure, *J. Chem. Phys.*, **36**, 2326 (1962).
99. H. Stegemeyer, *J. Phys. Chem.*, **66**, 2555 (1962).
100. D. Schulte-Frohlinde, H. Blume and H. Güsten, *J. Phys. Chem.*, **66**, 2486 (1962).
101. S. Malkin and E. Fischer, *J. Phys. Chem.*, **68**, 1153 (1964).
102. G. S. Hammond and coworkers, *J. Am. Chem. Soc.*, **85**, 2515 (1963).
103. R. S. Mulliken and C. C. S. Roothaan, *Chem. Rev.*, **41**, 219 (1947).

104. J. Saltiel and G. S. Hammond, *J. Am. Chem. Soc.*, **85**, 2515 (1964).
105. J. A. Barltrop and B. Hesp, *Proc. Chem. Soc.*, **1964**, 195.
106. E. J. Bowen, *Adv. in Photochem.*, **1**, 23 (1963).
107. H. L. J. Bäckström and K. Sandros, *Acta Chem. Scand.*, **12**, 823 (1958).
108. F. D. Greene, S. L. Misrock and J. R. Wolfe, Jr., *J. Am. Chem. Soc.*, **77**, 3852 (1955).
109. D. E. Applequist, E. C. Friedrich and M. T. Rogers, *J. Am. Chem. Soc.*, **81**, 457 (1959).
110. D. E. Applequist and R. Searle, *J. Am. Chem. Soc.*, **86**, 1389 (1964).
111. E. A. Chandross and J. Ferguson, *J. Phys. Chem.*, **45**, 3554, 3564 (1966).
112. H. Kautsky and H. de Bruijn, *Naturwiss.*, **19**, 1943 (1931).
113. C. S. Foote and S. Wexler, *J. Am. Chem. Soc.*, **86**, 3880 (1964).
114. T. Wilson, *J. Am. Chem. Soc.*, **88**, 2898 (1966).
115. E. J. Corey and W. C. Taylor, *J. Am. Chem. Soc.*, **86**, 3881 (1964).
116. B. Stevens and B. E. Algar, *Chem. Phys. Lett.*, **1**, 58 (1967).
117. K. Tickle and F. Wilkinson, *Trans. Faraday Soc.*, **61**, 1981 (1965).
118. A. Berthoud and D. Porrett, *Helv. Chim. Acta*, **17**, 694 (1934).
119. J. L. Bolland and H. R. Cooper, *Proc. Roy. Soc.*, **A225**, 405 (1954).
120. C. F. Wells, *Trans. Faraday Soc.*, **57**, 1703, 1719 (1961).
121. J. N. Pitts, Jr., R. L. Letsinger, R. P. Taylor, J. M. Patterson, G. Rechentwald and R. B. Martin, *J. Am. Chem. Soc.*, **81**, 1068 (1959).
122. G. Porter and A. Beckett, *Trans. Faraday Soc.*, **59**, 2038 (1963).
123. G. Weber and F. W. J. Teale, *Trans. Faraday Soc.*, **53**, 646 (1957); *Trans. Faraday Soc.*, **54**, 640 (1958).
124. W. H. Melhuish, *J. Phys. Chem.*, **65**, 229 (1961).
125. G. Weber and F. W. J. Teale, *Trans. Faraday Soc.*, **53**, 646 (1957).
126. E. C. Lim and J. D. Laposa, *J. Chem. Phys.*, **41**, 3257 (1964).
127. R. E. Kellogg and R. P. Schwenker, *J. Chem. Phys.*, **41**, 2860 (1964).
128. D. S. McClure, *J. Chem. Phys.*, **17**, 905 (1949).
129. R. E. Kellogg and R. C. Bennett, *J. Chem. Phys.*, **41**, 3042 (1964).
130. A. Weller and W. Urban, *Angew. Chem.*, **66**, 336 (1954).
131. A. Weller, *Z. Electrochem.*, **61**, 956 (1957).
132. W. R. Ware, *J. Am. Chem. Soc.*, **83**, 4374 (1961).
133. R. G. Bennet, R. P. Schwenker and R. E. Kellogg, *J. Chem. Phys.*, **41**, 3040 (1964).
134. V. L. Ermolaev and E. B. Sveshnikova, *Soviet Phys. 'Doklady' (English Transl.)*, **8**, 373 (1963).
135. V. L. Ermolaev, *Soviet Phys. 'Doklady' (English Transl.)*, **6**, 600 (1962).
136. K. Sandros and H. L. J. Bäckström, *Acta Chem. Scand.*, **16**, 958 (1962).
137. G. Porter and F. Wilkinson, *Proc. Roy. Soc.*, **A264**, 1 (1961).
138. P. G. Bowers and G. Porter, *Proc. Roy. Soc.*, **A299**, 354 (1967).
139. H. Labhart, *Helv. Chim. Acta*, **47**, 2279 (1964).
140. T. V. Ivanova, P. I. Kudryashov and B. Ya. Sveshnikov, *Soviet Phys.'Doklady' (English Transl.)*, **6**, 407 (1961).
141. R. E. Kellogg and N. C. Wyeth, *J. Chem. Phys.*, **45**, 3157 (1966).
142. A. R. Horrocks, T. Medinger and F. Wilkinson, *Chem. Comm.*, **1965**, 452.
143. H. Knibbe and A. Weller, *Z. Physik. Chem. (Frankfurt)*, **56**, 99 (1967).

4

Energy Transfer in
Radiation Chemistry

Milton Burton, Koichi Funabashi, Robert R. Hentz, Peter K. Ludwig, John L. Magee and Asokendu Mozumder*

4.1 INTRODUCTION

In photochemistry the efficiency of a process is reported in terms of quantum yield (or quantum efficiency), the number of molecules converted or produced per photon absorbed; the number is not directly related to energy efficiency but is of direct theoretical significance. In radiation chemistry, by contrast, the efficiency is reported as 100 eV yield (i.e. *G*), the number of molecules converted or produced per 100 eV of energy absorbed[†]. This latter number is an energy efficiency; it has no direct theoretical significance. The choice of a non-theoretical expression for report of efficiency in radiation chemistry was not an historical accident. It was made deliberately to avoid the implication that any one kind of excitation or ionization may be responsible for the effects observed when a chemical system is subjected to high-energy irradiation.

One of the major concerns of radiation chemistry is to explain observed *G* values. Such values may range from the order of 0·01 in the case of crystalline ferrocene[1] to a number of the order of 10^6 in the case of a pure dry gaseous mixture of hydrogen and chlorine[2,‡]. More common values, for liquids, are in the range of 1 to 10. Exceptions are explained

* The authors are members of the staffs of the Chemistry Department and of the Radiation Laboratory of the University of Notre Dame. The Radiation Laboratory is operated under contract with the U.S. Atomic Energy Commission. This is U.S.A.E.C. Document COO-38-526, prepared in January 1967.

† In photochemistry, in the cases of fluorescence and phosphorescence, quantum yields of photons are sometimes reported. In radiation chemistry, similarly, *G* (photons) may be reported.

‡ The last line of Table I in Lind[2] gives data corresponding to $G(HCl) > 10^6$ under alpha-particle irradiation.

in a variety of ways; cf. Section 4.6. Chain reactions are suggested in the cases of very high yields. The idea of some type of energy transfer is invoked when a minor component is preferentially affected or, alternatively, when it decreases the yield from a major component of a mixture. Inherent stabilizing features are suggested when a substance proves unusually resistant to the effects of high-energy irradiation. The details of these generalized explanations are a major concern of the radiation chemist. An understanding of energy deposition and the subsequent processes is fundamental to an explanation of the observed effects—or lack of effects. This chapter is limited essentially to a discussion of the processes in condensed systems.

The primarily important effect precedent to a chemical process in radiation chemistry is some kind of excitation in a particular molecule or group of molecules. The primarily important physical effect involves either a momentum transfer, as from high-velocity neutrons, or an electronic excitation by a charged particle. In the latter case, the incident primary is not necessarily a charged particle. An x-ray or gamma-ray, for example, may yield a charged particle by a photoelectric effect, by Compton recoil or by pair production. However, we are not concerned with the special effects which may result from the interaction of high-energy x-rays or gamma-rays with elements of high atomic number. In the radiation chemical effects discussed here, interest begins with the effects of charged particles.

Deposition of energy from energetic charged particles and localization of that energy to produce a chemical effect are processes which are clearly acceptable as real. However, the assumption that the two processes are identical or even that they occur in the same locale cannot be accepted as axiomatic[3]. In the following sections we discuss the interaction of an energetic charged particle with real systems (in which constituent electrons are bound), the distributions of such interactions, the time scale of subsequent events, the mechanisms by which energy may be transferred from the site of energy deposition to a more remote site, and the evidence (in luminescence and in chemical effects) for such processes.

4.2 ENERGY TRANSFER FROM THE INCIDENT CHARGED PARTICLE

4.2.1 Preliminary Remarks

In radiation chemistry the present concern regarding effects of incident charged particles is almost exclusively confined to energy ranges such that electronic excitation of the matter which stops them dominates the energy-loss processes. For exceedingly fast charged particles, a part of the energy loss may occur in nuclear encounters, with resultant bremsstrahlung, pair production or even cascade showers. At present in radiation chemistry, the only experimental case involving very fast charged particles is that of the incident electron; in that case the ratio of losses in nuclear encounters and in electronic processes is approximately $EZ/800$ where E is the energy of the electron in MeV and Z is the charge on the nucleus[4]. However, nuclear encounters themselves also produce a certain number of lower energy electrons. As far as radiation chemistry is concerned, these electrons may be treated exactly like the incident ones but the geometry of final loss events is slightly different from that of a single lower-energy primary; i.e. the microscopic *patterns* of energy deposition are the same but their distribution in space depends on the energy of the incident electron.

Slow moving, positively charged particles have a good chance of capturing an electron from the medium. This charge-exchange process governs the slowing down of a positively charged particle toward the end of its path. For example, in media of low Z a proton of $\sim 100\,\text{keV}$ is as likely to have an electron bound to it in a $1s$ orbital (i.e. to be a hydrogen atom) as it is to be free. Thus, all positively charged particles are eventually reduced to neutral atoms and become thermalized by elastic encounters. Slow electrons, on the other hand, continue to lose energy by electronic excitation and ionization until their energy falls below the lowest excitation potential. Sub-excitation electrons must lose a significant part of their energy by exciting molecular vibrations. The details of the thermalization of sub-excitation electrons are not definitely established; different mechanisms may be operative under conditions existing in different media.

4.2.2 Theory of Stopping Power

The capacity of a medium to extract energy from a moving particle by slowing it down is conveniently defined as the differential rate of loss of energy per unit length of the track of the particle $(-\mathrm{d}E/\mathrm{d}x)$. Among radiation chemists, it is generally known as Linear Energy Transfer (LET). Theoretically the best understood case is that of a heavy, fast, charged particle penetrating a gas of atoms. In radiation chemistry, however, one has often to deal with light particles (e.g. electrons) and condensed media (e.g. water as well as other liquids and solids). The modifications in the basic stopping-power theory necessary to include these cases are known only in general terms. In consequence, in certain special situations, theoretical gaps prevent direct experimental test.

Bohr's Theory

An early semi-classical theory of stopping power of a fast charged particle is due to N. Bohr*,[5]. Bohr's theory involves an impact parameter defined as the distance of closest approach between the incident particle and a particular electron of the medium were there no interaction (see Figure 4.1). In a sense, the straight trajectory of the incident particle is

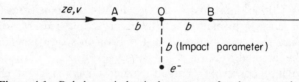

Figure 4.1   Bohr's semi-classical treatment for the energy loss
of a charged particle

equivalent to the Born approximation in quantum mechanics. With reference to Figure 4.1, the incident particle of charge ze and velocity v is considered as interacting only over the segment AOB of the path with (peak) force $-ze^2/b^2$ and for a duration of time $\sim 2b/v$. During this interval, the energy transferred from the primary particle† is given by

$$Q = (\text{momentum transfer})^2/2m = [(-ze^2/b^2)(2b/v)]^2/2m$$
$$= 2z^2e^4/(mb^2v^2) \tag{4.1}$$

The differential cross-section for this process for the range of impact parameters between b and $b-db$ is

$$d\sigma = -\pi d(b^2) = (2\pi z^2 e^4/mv^2)\cdot dQ/Q^2 \tag{4.2}$$

The stopping-power formula now follows from Equation (4.2) after summing over all the electrons which can be excited along a unit path and integrating over the relevant range of momentum transfer. Note that the differential cross-section obtained by Bohr applies strictly only to free

---

* Theories prior to Bohr, for example J. J. Thomson's theory, are only of historical interest.

† Here e and m denote the magnitude of the charge and the mass of the electron respectively.

electrons. For atomic electrons which are not free but can be excited*
with energy E_n, Bohr essentially surmised the rule

$$\sum_n f_n E_n = ZQ \tag{4.3}$$

where Z is the atomic number and f_n (later interpreted as oscillator
strength) is effective number of electrons in each atom which gains the
energy E_n. Substitution of the sum rule (4.3) into (4.2) gives

$$-dE/dx = (2\pi z^2 e^4/mv^2)NZ \int_{Q_{min}}^{Q_{max}} \frac{dQ}{Q} \tag{4.4}$$

where N is the density of the gas in units of number of atoms per cc. The
maximum energy transfer is given from the energy-momentum relationship as $Q_{max} = 2mv^2$. In the so-called impulse approximation, Bohr
obtained the minimum momentum transfer from the maximum value of
the impact parameter, b_{max}, given by the requirement that the collision be
sudden, i.e. $2b_{max}/v \leqslant \hbar/E_1$, where E_1 is a typical atomic transition energy.
Using the equality sign, Q_{min} is now given as $2z^2 e^4/mb_{max}^2 v^2 =$
$8z^2 e^4 E_1/(mv^2)(\hbar^2 v^2)$. With these values of Q_{max} and Q_{min} one gets from
(4.4)

$$-dE/dx = (4\pi e^4 z^2 N/mv^2) \cdot B \tag{4.5}$$

where B, the stopping number, is given in Bohr's theory by

$$Z \ln(2mv^2/E_1 \cdot \hbar v/4ze^2).$$

The impact parameter is not an observable. Consequently, the foundation of Bohr's classical theory is somewhat insecure. Despite this defect,
the theory has a wide range of validity and, if E_1 is appropriately defined,
Bohr's formula is practically indistinguishable from the quantum mechanical formula of Bethe except for an additional factor $\hbar v/4ze^2$ in the logarithm.
In the range of validity of Bohr's theory this factor is always much smaller
than the other factor in the logarithm, $2mv^2/E_1$, and is therefore of no
great consequence.

Bloch[6] has investigated the range of validity of Bohr's theory. Basing
his arguments on perturbation theory, he finds that Bohr's theory is
applicable if $v \gg v_e$ where v_e is the velocity of an electron in a 1s-orbit
around the incident particle. Usually this condition is easily fulfilled and,
as a matter of fact, it is less restrictive than the condition of applicability

---

* Unless otherwise specified, excitation includes ionization also.

of Bethe's quantum mechanical theory, which requires that $v \gg$ the velocity of atomic electrons of the medium*.

Bethe's Theory

The quantum mechanical stopping power theory due to Bethe†[7] considers energy and momentum transfers as basic variables. From elementary definitions, one can write a stopping-power formula

$$-dE/dx = N \int \varepsilon_n \, d\sigma_n \tag{4.6}$$

where $d\sigma_n$ is the differential cross-section for exciting the atom to a state n having energy ε_n above the ground state under impact by the charged particle. The interaction between the penetrating particle and the atomic electrons is, in general, electromagnetic; however, in the non-relativistic limit, it is well described by electrostatic interaction alone. In (4.6) as well as in other cases in this section, the integral sign implies summation over discrete states also.

The relevant loss processes can be broadly divided into two categories: (*i*) glancing collisions involving interactions at somewhat larger distances with a small energy loss per event and (*ii*) head-on collisions having interactions at closer distances with a higher energy loss per event. As is seen presently, approximately equal contributions to stopping power come from these two processes. The approximation employed here neglects relativistic effects; significant deviations may be expected when they are included[8].

The differential cross-section for glancing collisions in the range of energy loss ε to $\varepsilon + d\varepsilon$ is, according to Bethe,

$$d\sigma_g(\varepsilon) = \kappa \frac{f'(\varepsilon)}{\varepsilon} \left(\ln \frac{2mv^2}{\varepsilon} \right) d\varepsilon \tag{4.7}$$

where $\kappa = 2\pi z^2 e^4 / mv^2$ and $f'(\varepsilon) \, d\varepsilon$ is the dipole oscillator strength for excitation in the corresponding energy interval. Bethe obtains this closed formula by application of the Born approximation. The latter is the lowest non-vanishing term in the expansion of the expression for the interaction

---

* A possible exception, of course, is the case of a heavily charged positive ion traversing a medium of lower atomic number. However, in this case, charge exchange dominates the stopping and neither Bohr's nor Bethe's theory is applicable as such without modification.

† Details of Bethe's theory are quite involved; for a well-connected version of that theory see Fano[7(b)]. Here, however, we present a highly simplified treatment due to Magee[7(c)].

potential between the impinging particle and the atomic electrons. Further, the penetrating particle is heavy. Consequently, momentum transfer is much less than the initial momentum. This fact is employed in an additional approximation. Thus, the contribution to stopping power from glancing collisions derives from (4.7) as

$$-(dE/dx)_g = N \int \varepsilon \, d\sigma_g(\varepsilon) = \kappa N \left[(\ln 2mv^2) \int f'(\varepsilon) \, d\varepsilon - \int f'(\varepsilon)(\ln \varepsilon) \, d\varepsilon \right]$$

(4.8)

By definition, $\int f'(\varepsilon) \, d\varepsilon = Z$. Bethe defines a further quantity I called the 'mean excitation potential' by the relation

$$Z \ln I = \int f'(\varepsilon)(\ln \varepsilon) \, d\varepsilon$$

(4.9)

Substitution of these relations in (4.8) gives

$$-(dE/dx)_g = \kappa NZ \ln (2mv^2/I)$$

(4.10)

For the head-on collision process one can use the classical cross-section for each free electron, $\kappa \, d\varepsilon/\varepsilon^2$. The number of 'free electrons' is taken equal to the integrated oscillator strength for all transitions up to ε. Such usage is known as the 'dispersion approximation'; it gives

$$d\sigma_h = \kappa n(\varepsilon) \, d\varepsilon/\varepsilon^2; \qquad n(\varepsilon) = \int^\varepsilon f'(\rho) \, d\rho$$

(4.11)

The contribution to stopping power due to head-on collisions is now given from (4.11) as

$$-(dE/dx)_h = N \int \varepsilon \, d\sigma_h(\varepsilon) = \kappa N \int \frac{n(\varepsilon)}{\varepsilon} \, d\varepsilon$$

$$= \kappa N \left[(\ln \varepsilon) \int^\varepsilon f'(\rho) \, d\rho \big|^{\varepsilon = 2mv^2} - \int^{2mv^2} f'(\varepsilon)(\ln \varepsilon) \, d\varepsilon \right]$$

$$= \kappa NZ \ln (2mv^2/I)$$

(4.12)

The upper limit of the integral is the maximum energy transfer $2mv^2$. The small lower limit cancels from the integral terms and is therefore irrelevant. Here it is assumed that, while v is so large that essentially all oscillator strength is contained within $2mv^2$ and exact sum rules can be applied, nevertheless it is so small that relativistic effects are negligible. Within these limits, however, there is a wide useful range of energy. From Equations (4.10) and (4.12) it follows that glancing and head-on collisions contribute equally to the stopping power; in stopping-power theory this fact is called the *equipartition principle*. Thus, the total stopping power expressed in terms of the stopping number (cf. equation 4.5 et. seq.) is given by

$$B = Z \ln (2mv^2/I) \tag{4.13}$$

Discussion of Bethe's Theory

Bethe's theory necessarily inherits the limitations of the Born approximation. Normally it should require that v is much greater than velocities of atomic electrons. For the inner electrons for all atoms, except those in the first row of the periodic table, this condition is not easily fulfilled. The worst case is that of the K-electrons; however, they contribute insignificantly to the stopping power, and in many cases may be altogether omitted. On the other hand, there are special situations like K ionization, etc., in which they must be considered with appropriate correction indicated theoretically[6(b)] by Livingston and Bethe. However, this correction, as well as L-shell correction, whenever needed is best obtained by comparison with a few reliable experimental results[7(b)].

The Born approximation can be used properly only in the high velocity region. At low velocities only the Born perturbation approach[6(b)] is properly used. However, use of the Born approximation in that latter region gives approximately the same result. In the intermediate energy range the Born approximation is on an insecure foundation and special empirical procedures (such as shell correction, etc.) are necessary. For incident electrons this intermediate range is important, but many workers use the Born approximation anyway in absence of any other suitable simple theory. In the higher Born approximation, there is evidence of a correction to stopping power which is proportional to $(ze)^3$ and therefore depends on the sign of the charge. On the experimental side, as may be anticipated from this fact, there is some indication that the range of Σ^- heavy mesons is significantly larger than that of Σ^+ mesons generated from the same nuclear reaction[7(b)].

Bethe's assumption that all oscillator strength is contained within $2mv^2$ (cf. 4.12) is more inclusive than necessary. The result is that, for incident particles of lower energy, states are included which are energetically inaccessible. Consequently, Bethe's theory consistently overestimates the stopping power. In the case of protons penetrating a gas of H atoms, this overestimation is significant[9] at proton energies < 150 keV; however, in the same range of energies a more important contribution to stopping power is the consequence of electron capture from (and loss to) the medium. For heavier atoms, in which both charge exchange and failure of Bethe's approximation occur simultaneously at higher energies, the situation is similar.

The most important parameter in Bethe's formula is the mean excitation potential I, defined in (4.9). Its theoretical estimation requires knowledge of wave functions of all excited states. For the H atom, I can be calculated exactly; it is 15 eV. Using the Thomas–Fermi statistical model for the atom, Bloch[10] showed that $I \propto Z$. The constant of proportionality, called the Bloch constant, has been revised from time to time; at present a value between 13 and 14 eV is preferred. As may be expected from the fact that the Thomas–Fermi model is statistical, the conclusion that $I \propto Z$ fails at low Z. Even at high Z, exchange effects disturb the proportionality. In practice, the best estimation of mean excitation potential comes from comparison with experiment done in a reliable energy range, such as with protons of a few MeV energy. Figure 4.2 shows variation of (experimental) I/Z as a function of Z. A good part of this curve (the solid line) is explicable in terms of the Thomas–Fermi model including exchange; however, at present no theory explains the entire curve.

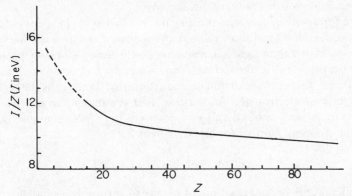

Figure 4.2 Variation of mean excitation potential as a function of atomic number according to Turner[11]. The Thomas–Fermi model gives $I \propto Z$. A more refined calculation including exchange gives $I/Z = a + bZ^{-2/3}$ with $a = 9.2$ and $b = 4.5$ as best statistically adjusted values

Modifications of Bethe's Theory

Elements of Bethe's theory presented so far are limited to the stopping of a fast, yet non-relativistic, heavy charged particle in a gas of atoms. Required modifications, when any of these conditions are not met, are indicated in the following paragraphs.

(*i*) *Molecules: The Bragg rule.* This important rule states that the stopping number, B, of a molecule is the sum of the stopping numbers of the constituent atoms. Because B depends logarithmically on I (Equation 4.13), the equivalent result for the mean excitation potential of a molecule containing A_1 atoms of atomic number Z_1 and mean excitation potential I_1, etc., is given by the law of geometrical averaging as

$$I \text{ (molecule)} = (I_1^{A_1 Z_1} I_2^{A_2 Z_2} ...)^{(A_1 Z_1 + A_2 Z_2 + ...)^{-1}} \tag{4.14}$$

For many compounds, Bragg's additivity rule holds impressively, with deviations of less than 2%. The reasons for the success of this rule are not entirely understood. However, we should note first that the contribution I_X of an atom X in various compounds is not expected to be the same as that of the free atom; that this contribution is nevertheless the same indicates a similarity of the chemical bonds in the various molecules. Furthermore, the excitations which involve most of the oscillator strength are high enough in energy so that specific details of chemical binding are of reduced importance. There is evidence that for lower incident-particle energies, Bragg's rule gradually breaks down.

(*ii*) *Light particle: Electron-stopping power.* When the incident particle is an electron, the maximum velocity of an ejected electron is somewhat less than v rather than the $2v$ characteristic of the case of a heavy particle. Furthermore, because the primary and secondary electrons are indistinguishable, they must be distinguished arbitrarily. Bethe calls the higher-energy ejected electron the primary, so that the maximum energy loss appears to be $\frac{1}{4}mv^2$ rather than $\frac{1}{2}mv^2$. Incorporating these changes, Bethe gives the stopping number for an electron as

$$B \text{ (electron)} = Z \ln \left[\frac{mv^2}{2I} \sqrt{(e/2)} \right] \tag{4.15}$$

The difference between Equations (4.13) and (4.15) is only a small factor in the logarithm; therefore, electrons and heavy (singly) charged particles with the same velocity in the non-relativistic range lose energy at about the same rate. This conclusion is not necessarily valid at relativistic energies nor at small incident energies.

(*iii*) *Relativistic effects*. Even if nuclear encounters are neglected, two modifications are required at high velocities. First, exact rather than approximate relations must be used for energy-momentum transfer, etc.; for example, the maximum energy transfer is $2mv^2/(1-\beta^2)$ where $\beta = v/c$ (rather than $2mv^2$). Second, at high velocities the interaction between the incident charged particle and the atomic electrons is not given accurately by static Coulomb interaction alone; at such velocities the electrodynamic interaction starts to be significant. In a certain representation*, although the energy levels to which an isolated atom (or molecule) may be excited by these two interactions may be the same (i.e. the state is degenerate), the states must differ in parity[7(b)]. Combining these two effects, the relativistic change in stopping number[7(a)] for heavy charged particles is given approximately by $-Z[\beta^2 + \ln(1-\beta^2)]$. For small β that expression is proportional to β^4 and so the importance of the relativistic effects is less than 0.1% for velocities less than 5×10^9 cm/s. For incident electrons, it is more convenient to write the full stopping number including relativistic effects[7(a)] as

$$B \text{ (electron)} = (Z/2)\left\{ \ln\left[\frac{mv^2 E}{2I^2(1-\beta^2)}\right] - [2\sqrt{(1-\beta^2)} - 1 + \beta^2]\ln 2 \right.$$
$$\left. + (1-\beta^2) + \tfrac{1}{8}[1 - \sqrt{(1-\beta^2)}]^2 \right\} \tag{4.16}$$

(*iv*) *Condensed phase*. A correction is needed for condensed material because in a condensed system, atoms (or molecules) are not isolated and their interactions are *correlated* to at least a distance $\sim\hbar/\Delta p$ where Δp is the momentum transfer† and because the interaction between incident particle and atomic electrons is somewhat reduced because of the *polarization* of the intervening medium. Fermi[12] first gave a theory to include the polarization effect. His theory is semi-classical and is based on the energy flow from the electromagnetic field set up by the moving charged particle in the dielectric medium. Because of the correlation condition,

---

* This is known as representation in the Coulomb gauge. In such representation the electrostatic interaction is instantaneous and force is exerted parallel to the direction of *momentum transfer*. The electrodynamic interaction is retarded and operates through the emission and reabsorption of virtual photons. The force so exerted is perpendicular to the direction of momentum transfer. Other representations are permissible but in the present situation the Coulomb gauge is convenient.

† For consistency within this chapter Δp is employed for momentum change or momentum transfer. However, readers who refer to much of the original literature will discover that q has been employed fairly consistently for the latter term. It may further trouble the reader to discover in this section that the *subscript p* denotes plasma!

the density effect is more important for low Δp collisions (glancing); this situation is true for both small and large incident velocities. Because of the polarization condition noted, the density effect actually prevents the relativistic stopping power from going to infinity as $\beta \to 1$ (see the previous paragraph) but instead allows it to rise gradually to a plateau called the Fermi plateau. The density effect is usually incorporated in the stopping number by adding a corrective factor $-Z\delta/2$ where δ is called the density correction. Fano[7(b)] gives an expression for δ in terms of the complex dielectric constant which for high incident velocities simplifies to $\delta \to \ln [\hbar\omega_p{}^2/I^2(1 - \beta^2)] - 1$ where $\omega_p{}^2 = 4\pi Ne^2 Z/m$.

When the condensed medium is in the form of a thin film with vacuum on either side, the energy loss is frequently seen to occur in the electronic excitation of collective (i.e. plasma) oscillations. Either one or several quanta may be excited at once. Typical loss quanta lie in the range 10–25 eV and are substantially independent of atomic or molecular characteristics[13]. Out of several possible fates of the lost energy, a significant one, actually observed, is the emission of radiation (Ferrell radiation)[14]. No chemistry has been reported as caused by this particular loss process in thin films; the effects are weak, if any, and have not yet engaged the attention of chemists.

4.2.3   Further Consequences of Stopping-Power Theory

There are many ramifications to stopping-power theory and its applications. Of these, a few, enumerated in this section, are believed to be of importance to radiation chemistry.

Range and Straggling

The distance, measured along the actual path, that an incident particle travels before it is brought to rest is defined as its range. A sufficiently fast particle travels more or less in a straight line even though there may be significant scattering in high-energy head-on collisions (δ-rays) as well as in bending (at low velocities) near the end of the trajectory. Penetration measures the projection of the path along the direction of original motion[*,15]. Slower electrons show considerable scattering.

Range may be obtained by (numerical) integration of a suitable stopping-power formula (which includes all the necessary corrections;

* The maximum observed penetration of particles of a particular energy is the same as its range.

see Section 4.2.2). A stopping-power formula gives only the average LET; because of statistical factors, individual particles suffer unequal losses under similar conditions. In fact, the collision process itself is statistical. The effects of inequality in the loss processes on range are reflected in straggling. There are basically two kinds of straggling: (*i*) energy straggling describing distribution in energy of particles starting with the same energy and travelling a fixed distance under identical conditions and (*ii*) path-length straggling describing distribution in travel distance between fixed energies. Both stragglings may be obtained from the same basic equation[7(b)]; however, in radiation chemistry the important quantity is range straggling* (i.e. path-length straggling at final energy zero). Under trivial approximations always satisfied in radiation chemistry, range straggling exhibits a Gaussian distribution; i.e. the probability of having a range R is given by $(2\pi \overline{\Delta R^2})^{-1/2} \exp[-(R-\overline{R})^2/2\overline{\Delta R^2}]$ where $\overline{R}$ is the mean range and $\overline{\Delta R^2}$ is its mean square dispersion. In actual practice range straggling is quite considerable; a typical value of $(\overline{\Delta R^2})^{1/2}/\overline{R}$ is $\sim\frac{1}{4}$.

For very fast incident particles, the velocity dependence in the logarithm of Bethe's stopping power formula (cf. Equations 4.5 and 4.13) may be neglected in comparison with the v^{-2} term in front of the logarithm; as a result, the range tends to be proportional to the square of the energy. This condition is found to be approximately true for most incident particles at high energies. At lower energies the logarithmic term (as well as charge exchange for positive particles) may become important and range depends on a smaller power of energy. For electrons in water[8] in the energy interval of a few thousand to a few hundred eV, for example, the power of the energy term progressively diminishes from 1·75 to 0·9. Typical computed values of ranges in water are: (*i*) electrons of 5 keV energy, 5600 Å, (*ii*) protons of 5 MeV energy, 0·5 mm, and (*iii*) α particles of 5 MeV energy, 0·04 mm. Ranges of fission fragments and heavy charged particles are almost proportional to energy; e.g. $\sim$40 mg/cm^2 in Al for Ne ions of energy $\sim$200 MeV (i.e. for ions with energy 10 MeV per nucleon). The reason for this near proportionality is the fact that the increase in stopping power as the energy of the particle diminishes is almost completely neutralized by decrease of effective charge on the ion resultant from electron capture. The range-energy relationship is of importance to the

---

* A simple concept for the range straggling is the following. Glancing losses are numerous and each is quite small; hence their effect is fairly the same for all particles. However, the head-on losses are few and each one is quite large. Therefore, the particle which has even a few more head-on encounters is brought to rest in a significantly less distance compared to others.

radiation chemist. In addition, it provides the experimental physicist with a tool to measure particle energy or even to identify the particle. Numerous range-energy graphs and tables are at present available; a recent publication of the National Academy of Sciences—National Research Council, U.S.A. (publication No. 1133) is particularly noteworthy[11].

Generic Effects

It is stated in Section 4.2.2 that energy transfer from the primary particle divides more or less equally between glancing and head-on collisions. Head-on collisions often give rise to ionization with a secondary electron distribution in number varying almost inversely with the square of energy (see Equation 4.11). These secondary electrons produce tertiaries, etc., until no newer generation is energetically possible. By virtue of lower energy, the later-generation electrons have much higher LET compared to those of the earlier generation. On the contrary, for a high LET primary, the secondary electron may be of lower LET. In any case, because some important chemistry (e.g. the 'molecular yield' in water[16]) depends strongly on LET, generic effects may be significant[8,17].

Capture and Loss*

An ion with a high positive charge z (e.g. the bare nucleus) having velocity v much greater than that of an electron in K-orbit around the ion $(v' = zh/e^2)$ will lose energy in induction of electronic excitation, etc., on entering a medium. The charge on the ion remains essentially constant until the velocity is reduced to $\sim v'$, whereupon the probability of capturing an electron from the medium becomes significantly large. However, the captured electron is almost invariably lost in the next few electronic collisions. In the beginning, this interplay of capture and loss is characterized by relatively large cross-section for loss compared to that of capture. Meanwhile, as the particle is slowed down, capture probability increases rapidly at the expense of loss probability and, at a certain stage, an electron is permanently bound in the 1s orbit around the ion. The ion now travels with charge $z - 1$ and essentially its previous history is repeated; i.e. it captures and loses a second electron while being slowed down and eventually also binds another electron permanently. The sequence of events continues until the ion is reduced to a neutral atom,

---

* Only a brief account is presented here. Recently, details of energy loss processes of heavily charged ions have been reviewed by Northcliffe[18].

whereupon it loses its final kinetic energy by elastic encounters and is eventually thermalized.

Because of the charge-exchange processes described in the preceding paragraph, the stopping power tends to decrease as the ion slows down (LET $\propto z^2$). However, the slowing down process itself tends to increase the stopping power (LET is roughly $\propto v^{-2}$, neglecting the slowly varying logarithmic term). As a result of these two counter-effects, the stopping power at first increases slowly, then goes through a rather broad maximum (reflecting approximate cancellation of the two effects) and finally decreases when the charge is significantly reduced*. After electrons have been captured into the two 1s states and under certain conditions, the intermediate region of more or less constant LET may become of dominant importance in the sense that most of the ion energy is lost in this region. If this situation is indeed true, the range is almost proportional to energy. Such conditions are usually encountered for ion energies of a few MeV per nucleon traversing a medium of low atomic number.

Because capture and loss processes are statistical, detailed examination of the cross-section of a monoenergetic beam of heavy ions passing through any medium would reveal particles in all charge states ranging from the bare nucleus value to the neutral atom. The velocity of the beam (in relation to the velocity of a 1s electron around the ion) is the important clue to the probability distribution; the conclusion from such a consideration is that the charge distribution is roughly independent of the nature of the impinging ion[18,†]. The distribution itself is very sensitive to velocity and charges far from the average value are highly improbable. Thus for C^{+6} ions in aluminum at a velocity of 1.3×10^9 cm/s (equal to 1s orbital velocity; ~ 0.9 MeV/nucleon), the probabilities of $+6$, $+5$, $+4$ and $+3$ charges are respectively 0·23, 0·49, 0·26 and 0·02, giving an average charge of 4·93. The same values at only a somewhat higher velocity of 2×10^9 cm/s (~ 2 MeV/nucleon) are respectively 0·62, 0·33, 0·05 and 0·00 giving an average of 5·57. At the higher velocity in this case (2×10^9 cm/s), maximum charge contribution comes from the bare nucleus, while at the lower (1.3×10^9 cm/s), the state with one electron captured contributes most to the charge of the beam.

---

* The statement refers to electronic stopping power only. There is a further maximum at low ion-velocities in the stopping power of heavily charged ions attributable to elastic nuclear collisions. See Figure 2 in Northcliffe[18].

† Of course, the maximum charge *is* affected by the nature of the impinging ion. The import of the statement is qualitative.

Electron-stopping Power

Usually, radiation of a given quality has a characteristic range of LET. In water, the LET values of protons and α-particles of a few MeV energy, of heavy ions around a few MeV/nucleon and of fission fragments are respectively around a few eV/Å, a few tens of eV/Å and a few hundreds of eV/Å. However, electron-stopping power may vary greatly with energy. In water an electron beam of ~ 1 MeV or the Compton recoils from ^{60}Co gamma radiation may have a small LET ≈ 0.02 eV/Å while secondary electrons generating short-length tracks may have a high LET ≈ 1 eV/Å. Electron-stopping power is important in radiation chemistry because the electron (whether as a γ-ray Compton-recoil electron or as a secondary) is the important producer of initial effects; in any case, a substantial amount of energy appears in the secondary electrons irrespective of radiation quality. Figure 4.3 shows electron-stopping power in hydrogen as a function of energy. With increase in energy, the stopping power first decreases because of the v^{-2} term; the relativistic rise begins at ~ 2 MeV. Finally the stopping power rides to the Fermi plateau at ~ 100 MeV because of polarization screening (density effect). The relativistic rise has been observed only in the case of the electron[9] but will be observed for protons in the accelerators under present development.

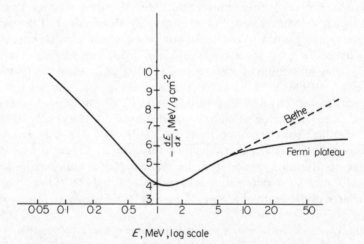

Figure 4.3 Electron-stopping power in hydrogen as a function of energy. Note the uniform decrease in the non-relativistic region, the relativistic rise and the gradual ascent to the Fermi plateau

4.2.4 Geometrical Effects in Energy Transfer from the Incident Particle[19,20,21]

A charged particle produces a track in any medium it traverses. This fact means that the loss events are geometrically correlated. Tracks of massive charged particles and fast electrons conform more or less to straight-line geometry. Slow electrons show considerable scattering. From certain viewpoints (such as radiation chemistry, the operation of particle detectors, etc.) it is important to know whether the energy lost by the particle remains in the vicinity of the track for a reasonable time duration. Energy lost in the form of fast secondary electrons, Cerenkov radiation or bremsstrahlung is removed from the track within times $\gtrsim 10^{-15}$ s. Such losses are known as unrestricted losses. In radiation chemistry, up to the present time, the role of unrestricted losses has been negligible because (i) energy lost in bremsstrahlung and Cerenkov processes does not exceed a fraction of a percent at most, and (ii) energy lost in the form of a fast secondary electron may be incorporated in the calculation as restricted loss of an incident particle in the track that it now produces. Under these conditions, the LET and the stopping power become equal (cf. Equations 4.10, 4.12 and 4.13).

In physics, particle tracks are traditional devices for particle detection and even for measurement of flux (as in counters), energy and velocity. Studies of tracks in the cloud chamber led ultimately to the bubble chamber with numerous counters and solid-state devices along the way. In the radiation chemistry of water particularly, an explanation of yields of radical and 'molecular' products as well as of solvated electrons requires track models. In fact, diffusion kinetics as applied to radiation chemistry starts with the hypothesis that chemical changes brought about by radiation can be explained as reactions, of primarily produced species with each other or with reactive secondaries or with solvent or solute molecules, along the track of the particle. At present, it can be said that diffusion kinetics has good qualitative and semi-quantitative application. Other phenomena involving high-energy radiation effects that may require track models include certain aspects of luminescence, polymerization, protection, sensitization, etc.

High LET particles, such as α-particles and protons, produce dense columnar ionizations resulting essentially in cylindrical tracks. A low LET particle, such as a fast electron, produces ionization and excitation centers separated by great distances. These centers lie more or less on a straight line which defines the track skeleton. Thus we have the model of a 'string of beads' with the beads as isolated inelastic loss events, the string being the track. The length of a track is the range of the primary particle. The diameter of the track for high LET particles or the size of 'beads' in the low LET track depends on the typical distance within which the lost energy may be supposed to be confined and ultimately made available for chemical reaction or energy dissipation. Frequently this distance is taken to be ~ 10 to 20 Å. In water, a typical 'bead' contains an energy ≈ 40 eV and a typical α-track contains $\sim 10^7$ ions/cm initially. Tracks of different LET have different structures.

Spurs

In radiation chemistry, there is a real separation in time scale between the physical state, the onset of which is marked by energy deposition ($\sim 10^{-16}$ s), and the chemical stage which is heralded by significant chemical reaction. Certain ion–molecule reactions (e.g. charge transfer) can be faster than other reactions; nevertheless, it can be said that the chemical stage lags the physical stage by orders of magnitude in time (cf. Section 4.3). In this intervening physico-chemical stage, therefore, there is time for energy cascading of excited electronic states, for energy degradation, etc. Much later, thermalized radicals and ions are available for chemical reaction either among themselves or with a solute that may be present. The primary species of particles (e.g. radicals or ions) suffer numerous collisions in different directions in the process of thermalization. The result is a more or less spherical region which expands by diffusion and in which most of the chemical reactions involving primary species are confined. In the quantitative theory of track reactions presented by Samuel and Magee[19], they call these spherical regions or beads 'spurs' and were non-committal about the response of the matter itself. According to their theory, spurs are widely separated (several thousands of Ångströms) on fast-electron tracks in water so that chemical reactions develop essentially independently in each spur. Secondary electrons of high energy produce tracks similar to the primary (but with much larger LET). In high LET tracks (protons, α-particles, etc.) the spurs overlap to give cylindrical tracks. The first accomplishment of the Samuel–Magee theory was the successful computation of relative radical and molecular yields in water irradiated by ^{60}Co gammas, ^{3}H betas, and 6 MeV alphas. In that theory, however, there is no device for handling the consequences of the inherent slowing-down of the incident particle. This difficulty was removed by Ganguly and Magee[20] by considering the structure of a low LET track as a random succession of spurs, the average spacing of which conforms to the local LET in that part of the track. Slowing down is inherent in such a model and spur overlap is treated as the consequence of spur expansion by diffusion.

Spurs, Blobs and Short Tracks

Both the Samuel–Magee and Ganguly–Magee theories have been successful in many areas of radiation chemistry and their utility for contemporary experimental radiation chemists has been considerable. However, they are not free from obvious physical shortcomings. For example it may be pointed out that a low-energy secondary electron having energy $\sim 100\,eV$ has, according to the Onsager theory[22], an exceedingly poor probability of escaping recombination with its sibling positive ion. Recombination results in an extended region in which much energy is locally available. In a more refined track model, these entities have been called 'blobs' by Mozumder and Magee[8,17,21]. According to their theory a low LET track is better described in terms of three recognizably different entities called spurs, blobs and short tracks. They are defined as follows.

Spurs are the consequences of all energy losses in which the effects of oscillator strength dominate (except for the small contribution—perhaps amounting to 2%—which may be attributed to K-electron excitation). For molecules containing atoms from the first two rows of the periodic table, a convenient limit of energy loss may be taken as $\sim 100\,eV$. Spurs may be supposed to be generated by glancing collisions whereas all extra-spur entities originate from head-on processes.

Blobs result from secondary electrons generated in a knock-on process ($> 100\,eV$) produced, however, with energy insufficient for immediate escape from the parent ion. Onsager gives the escape probability for a pair of ions as $\exp(-e^2/DRkT)$ where R is the range of the secondary electron and e, D, k and T are respectively electronic charge, short-time dielectric constant, Boltzmann constant and absolute temperature. Combining this probability with the (extrapolated) range-energy relation for a low energy electron in water, Mozumder and Magee obtain an upper limit of $500\,eV$ for the energy content of blobs in that medium.

Short tracks result from secondary electrons sufficiently energetic to escape parent-ion recombination but yet with energy so low that successive loss events overlap to a great degree. A defined upper limit, so to speak, of $5000\,eV$ is obtained for short tracks by equating the mean free path between inelastic collisions to ten times the spur radius. For δ-rays of energy higher than 5000 eV, the tracks are called *branch tracks*; the loss events resultant from glancing may be supposed to be essentially isolated in them. Short tracks (and blobs) represent local high LET sections of the overall low LET track; their effect on chemical yields roughly corresponds to the fraction of energy deposited in the short track form.

Figure 4.4 shows, as a function of incident energy, the way that energy in an electron track in water is partitioned in the three categories of spurs, blobs and short tracks. In this consideration the branch tracks are not separately categorized but are decomposed into their constituents (i.e. spurs, etc.); the process is continued until no more branch tracks are energetically possible. The demarcations in energy between the three categories are represented only roughly for they may depend on the medium. In any case, it is profitable to separate the entire track into regions of roughly low and high LET because competitive chemical reactions are affected quite differently in the two ranges.

4.3 THEORETICAL CONCEPTS OF ENERGY FLOW IN TRACKS

4.3.1 Some General Notions

The response of matter to the energy added by a fast charged particle is much less well understood than is the stopping power (cf. Section 4.2)*. In the latter case a basic theory exists and direct checks by experiment are possible; one merely measures the ranges of particles in matter, a relatively easy task. The pattern of response of the medium presents an *essentially* more difficult problem. Experimental measurements pertain to processes later than the primary events, which are of immediate concern. Indeed, it is not known whether the primary events are actually measurable. Measurements are made on ultimate phenomena, not all of the same quality, such as emission of visible light, formation of intermediates which absorb light, and yields of isolable chemical products. Frequently, many generations of processes occur between the absorption of energy and the measurement of some phenomenon of interest to an experimenter, and the connections between the absorption of energy and the event studied are difficult to disentangle.

---

* The various processes resulting from the interaction of an impingent particle with matter are significantly affected by the degree of aggregation of the medium traversed. It is important to emphasize that this entire treatment refers to the behaviour of condensed states.

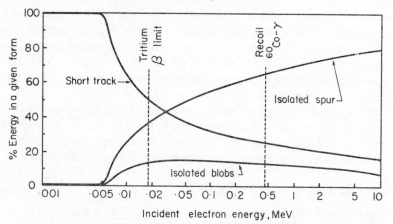

Figure 4.4 Graphical plot of percentage energy split between spurs, blobs and short tracks for electrons in water, according to Mozumder and Magee[8]

It is unlikely that a general theory, in the sense of stopping power theory, can even be developed to describe the response picture. On the other hand, some general notions are useful; they provide a speculative framework in this area. We attempt to examine them here.

The loss of energy by a fast charged particle as it traverses a medium is accompanied by the production of excited states in that medium. Historically, the earliest description of such a state was that all the energy was given to one electron which was ejected as a recoil electron and left its parent molecule in the ionized form. For energy losses above 100 eV this description of the excited state is almost certainly adequate; for energy losses in the 10–30 eV range, which constitute the largest class of energy dissipation (i.e. in 'spurs'; cf. Section 4.2), it is certainly not adequate. In any case, the high electronic excitation is reduced by sharing the kinetic energy with other electrons until only the lowest electronic excited states of the medium remain. Details of the downward 'cascading' of excitation depend upon the nature of the medium and there is evidence that much variation in the yields of lower excited states is possible[*,23].

On the most general grounds it is expected that a significant fraction of the absorbed energy is converted into low-grade heat even at a fairly early

* In the cases of benzene and cyclohexane as examples, the former may end up in a variety of singlet and triplet states which can internally convert (or 'intersystem cross') to each other. In cyclohexane the immediate consequence (i.e., in $\sim 10^{-13}$ s) of production of *any* excited species other than an ion appears to be decomposition[23].

time (i.e. within 10^{-13} s). Of course, as the absorbed energy is 'thermalized', high-energy intermediates, such as ions, radicals, etc., are formed. The excess energy of these intermediates is electronic although it is not customary to call such species excited electronic states. Exothermal reactions of the intermediates continue the conversion of the originally deposited energy into low-grade heat.

An important feature of radiation chemistry is the correlation of positions of formation of reactive intermediates in the path of the incident particle. Track effects arise from this correlation because the intermediates react with each other or with the medium in competitive processes.

4.3.2   The Time Scale of Events

Many orders of magnitude in time are involved in the course of the various responses in matter traversed by the species which donates the energy[24]. The shortest distinguishable time interval is obtained from the uncertainty relationship of quantum mechanics (Heisenberg principle)

$$\Delta t \approx \frac{\hbar}{\Delta E}$$

Because a large part of our concern is with spurs which involve energy losses of around 30 eV,

$$\Delta t \approx \frac{6 \times 10^{-16}}{30} \approx 2 \times 10^{-17} \text{ s}$$

and the shortest time of significance in radiation chemistry is of that magnitude; it is the time interval required for the usual energy deposition act.

The transit time of a 30 eV electron between molecules separated by ~ 4 Å between centers is approximately 10^{-16} s. Low-energy electrons must lose any excess energy they have greater than that required to produce electronic excitation in the medium, in a few molecular collisions[25]. The time required for an electron to fall into the sub-excitation region of energy may require about 10 collisions and is thus expected to be of the magnitude of 10^{-15} s.

The initial time interval of $\sim 10^{-15}$ s, during which the expelled electron occupies the center of attention, is too short for nuclei to move or for molecules to be dissociated or even vibrationally excited. However, this time interval is sufficiently long so that energy can be transferred between electronic states of the medium. It is notable that the uncertainty principle requires that the states involved in such a process cannot be established more exactly than 1 eV.

The next time interval of significance is set by the vibration time of molecules ($\geqslant 10^{-14}$ s). In that time interval, a large fraction of the absorbed energy can be transferred to molecular motions. The mechanism of excitation of vibrations is not momentum transfer from the electrons to the nuclei in collisions; it arises from the Franck–Condon effect. Any electronic excitation changes the vibrational state of a molecule. A spur has a complicated electronic excited state which consists of several excited molecules, ions and free electrons; the relaxation of the medium naturally involves much vibrational excitation.

On a more detailed mechanistic level it can be said that much of the vibrational excitation arises from low-energy electron impacts on neutral molecules. It is now well known[26] that such excitation processes occur through intermediate negative ion states and usually have large cross-sections. Although much is known about this process for gas phase molecules, little is known concerning the condensed phase. However, there is little doubt that an intermediate electron capture process is effective.

In the next time interval, of $\sim 10^{-12}$ s, vibrational energy flows into other modes of motion and gross movements of molecules become important. In condensed media they include rotations, intermolecular vibrations and the beginning of relaxation phenomena. In that time interval, the term 'local temperature' begins to have real significance*.

Figure 4.5 shows in a schematic way how the energy deposited in a spur is transformed in time.

After a local temperature is established, a large fraction of the energy has been transformed into low grade heat. The amount which remains in electronic excitation (excitons) and high energy intermediates depends upon the specific properties of the medium. In any case, however, significant increases in local temperature are not expected to be produced by low LET radiations; i.e. those radiations the primary track of which produces widely separated spurs. In such case, the secondaries do produce more concentrated conditions of energy release but the total amount of energy involved in each such secondary track is relatively small.

---

* These statements emphasize a significant difference between radiation chemistry and conventional reaction kinetics. In shock-wave phenomena as an example, energy becomes successively more and more localized. Mechanical energy is transformed successively into molecular translations, vibrations and into electronic excitations.

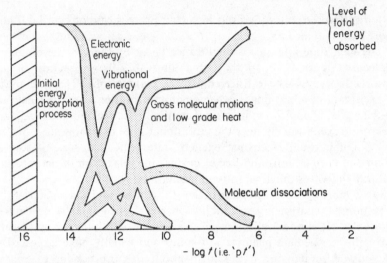

Figure 4.5 Schematic description of the partition of energy in a spur (cf. Section 4.2) as a function of time t, in seconds. The vertical dimension of a given region is proportional to the fraction of the deposited energy which is in a particular form. This figure is intended to describe the situation for a spur having average energy. Note that, in exothermal reactions, the line showing low-grade heat can terminate above the level of total energy absorbed. This situation is the same even when luminescence occurs, because the amount of heat loss in such processes is rarely significant

The reactions within track entities which require diffusion of molecular intermediates require much longer times, 10^{-8} to 10^{-6} s. The instantaneous concentration of tracks depends upon the dose rate (rate of irradiation) as well as on such specific properties of the medium (e.g. viscosity, diffusion coefficient of intermediates, decay times of excited states) as affect the disappearance of the tracks. Thus, the intermingling of intermediates formed in different tracks has a time scale which depends upon the rate of irradiation, but is usually of a considerably longer time scale than that of reactions within an isolated track.

4.3.3 Fate of Electronic Excitations

In terms of elementary, but qualitative, considerations[27] stationary electronic states of a condensed system have properties which are in the general category of wave packet states. The energy is *initially* deposited in one molecule or the few molecules immediately surrounding, i.e. the excited state is created by a perturbation with a limited range of inter-action. The initial excitation must be in that range. However, that state itself must be recognized as a superposition of stationary states which initially have destructive interference everywhere except in the region where the excitation is found. At later times (as short as 10^{-15} s; cf. Section 4.3.2) the interference pattern changes and the region of excitation spreads.

The initial condition for a packet state comes from the uncertainty relationship. Such a state cannot be localized more precisely than is given by the de Broglie wavelength associated with the momentum transfer Δp from the particle, i.e.

$$\lambda = \frac{h}{\Delta p} \approx \frac{hv}{\varepsilon} \tag{4.17}$$

where ε is the energy given the matter.

For an energy loss of 15 eV, $\lambda \approx 900$ Å if the energy of the incident electron is 50 keV. This result is very general and does not depend upon detailed properties of the system, its state of aggregation or its chemical composition. Such a state is called a 'delocalized' state. If the coupling of the system is strong, the excited state is a broad band of energies and spreading of the packet is rapid. However, if the coupling is weak, the excitation is essentially confined to a single molecule. In the latter case the large size of the packet is merely related to the uncertainty in the location of the excited molecule.

The concept of 'localization' of energy following formation of a de-localized excitation has been much discussed. Such a process is of interest in connection with several problems. For example, delocalized states in multi-component systems may preferentially yield an effect, such as a dissociation, in one component. Preferential effects are actually quite common but it has never been demonstrated that they actually originate at this early a stage.

In the radiation chemistry of low LET radiation, most track entities involve energy losses of such magnitude (i.e. 20–40 eV) that conversion processes into successively more numerous excitation states must be rapid. At the end of the time interval associated with electronic excitation (cf. Figure 4.5), several excitons involving only the lowest molecular excited states remain.

It is the *correlation* of intermediates which is important. From the nature of the interaction (they are all very short range) it is clear that ionizations and dissociations must occur quite close together. Thus, one understands in a general way the possibility that a spur initially delocalized over perhaps as much as 900 Å yields a highly excited region of limited extent. In this way, most of the energy is localized so that the next process involves only one molecule or its nearest neighbors. Consequently, the set of molecular (or free radical) intermediates which result are confined to separations of 10–20 Å.

4.3.4 Flow of Low-Grade Heat

The magnitude of the temperature increase at times when it is appropriate to use the concept of local temperature is of interest. In the vicinity of a track entity which has spherical symmetry (spur, blob) the temperature distribution should have the approximate form*

$$T = T_a + \Delta T \frac{\exp\left[-r^2/(r_0^2 + 4Dt)\right]}{(1 + 4Dt/r_0^2)^{3/2}} \tag{4.18}$$

where T = the temperature at distance r from the center of the spur, T_a = the ambient temperature, ΔT is a temperature increase parameter, and r_0 is a spur-size parameter. The macroscopic values suggest that $D \approx 10^{-3}$. For water, with $C_v \sim 4 \times 10^7$ erg/cm^3 degree, a spur with $r_0 = 20$ Å[†,28] will have an initial temperature at its center of 30° above the ambient‡. The time for this temperature to drop to one-half its value is just slightly less than 10^{-11} s§.

This temperature increase is too small and its duration too short to produce purely thermal reactions. Consider a radical reaction with activation energy 8 kcal. If its rate constant for some reaction with the

---

* This is a typical Gaussian expression for the excess temperature. At the earliest time at which it is appropriate to speak of temperature, the distribution is expected to be of this type because the processes involved are wholly random. Furthermore, any initially asymmetric distribution may be expected to relax into such a symmetric form in a time of the order of $r_0^2/4D$.

† This is the general magnitude of spur radii used in diffusion kinetics.

‡ This temperature is obtained from the relation

$$\varepsilon = C_v \Delta T \int_0^\infty \exp\left(-\frac{r^2}{r_0^2}\right) 4\pi r^2 \, dr = \pi^{3/2} r_0^3 D C_v T$$

with $r_0 = 20$Å, $\varepsilon = 30$ eV.

§ The half-time for the temperature drop at the center is obtained from the relation $[1 + (4Dt/r_0^2)]^{3/2} = 2$. With the parameters mentioned in this discussion, $t_{1/2} = 0.6 \times 10^{-11}$ s.

substrate is

$$k_R \approx 10^{-11} \exp\left(-8000/RT\right) \text{cm}^3 \text{s}^{-1} \qquad (4.19)$$

we have that

$$\frac{\mathrm{d}P}{\mathrm{d}t} = -Pk_R C \qquad (4.20)$$

where P is any representation of the concentration of free radicals and C is the concentration of molecules available for reaction and may be as large as 10^{22} molecules cm^{-3}. Even if the temperature were as high as 400°K, we would have

$$-\frac{\mathrm{d}\ln P}{\mathrm{d}t} = 10^{11} \, e^{-10} \approx 2 \times 10^6 \qquad (4.21)$$

so that the reaction time is 5×10^{-7} s. Thus it follows that a thermal pulse which lasts 10^{-11} s is inadequate to make the reaction go significantly even with ΔT of 100°.

Chemical intermediates formed in the sequence of events following a single energy absorption process are initially in the vicinity of the absorption. As they diffuse away from their birthplaces they may undergo mutual reaction in a typical 'track process'. The diffusion of sibling intermediates will ultimately cause them to mingle with unreacted intermediates from other spurs or tracks. This part of the reaction is more typical of homogeneous kinetics and little if any effect of the special character of the radiation origin of the intermediates remains.

4.4 MIGRATION OF ELECTRONIC ENERGY: SPECIAL EFFECTS IN TRACKS

4.4.1 General Nature of the Energy Transfer Problem

Much of our knowledge of electronic energy transfer comes from studies of systems excited photochemically. In such cases the electronic states of the molecules involved in the migration of electronic energy are usually well known, as are their multiplicities, symmetries and excitation energies. In general, the processes of excitation of these are optically allowed and the excitation energy is somewhere between 3 and 10 eV. In the radiation chemistry of condensed systems the relevant electronic states include not only such low-lying triplet states as may be produced in optically forbidden processes, but also highly excited states such as super-excited states[29] or plasmon states[30].

A super-excited state may represent a state of a single molecule in a gas; the conditions in a condensed system are, however, such that the actual excitation may overlie several molecules. Thus, the state itself may more properly be assigned to a part of the system rather than a molecule. Nevertheless, the ultimate effect associated with the very rapid disappearance of the state (e.g. autoionization) does occur in a single molecule.

A plasmon state is a special case of an exciton state of extremely short life; its energy belongs to a number of molecules—not to anyone individually except, perhaps, in the process of an ultimate effect. Although the lifetimes of plasmon states are generally held to be very short, the number of molecules which their excitation overlies may be so great that energy 'migration' appears to be extremely rapid and, with disappearance of the state, any one of these molecules can be, with some probability, chemically or physically affected.

Another class of electronic states must also be considered in the case of radiation chemistry of the condensed phase and particularly of rigid media. These states can be called charge-transfer or charge-separated states and their energies are below the first ionization potential of the isolated molecule (when extracted from such a system). When a charge-separated state migrates in a crystal with a well-defined momentum, it is called a 'Wannier' exciton, whereas the migration of neutral excited states corresponds to a 'Frenkel' exciton[31].

The relationship between the neutral excited states and the charge-separated states is describable in the following terms. When the excitation energy is not too large, the electron in the excited orbital is able to complete its periodic motion with only slight disturbance by the neighboring molecules. In this case the behavior of the excited molecule is very similar to that of the excited molecule in the gas phase; i.e. to that of a neutral excited state. At higher excitation energy, the orbital motion of the excited electron is perturbed by the surrounding molecules. When the frequency of disturbance is so high that the electron cannot complete even one periodic cycle, the electron no longer belongs to the original molecule alone, and the behaviour of the excited state can be quite different from that of the excited molecule in the gas phase. If there is a mechanism of trapping such an excited electron either in the form of a negative ion or in that of a 'solvated' electron, charge-transfer or charge-separated states can be formed.

Charge-separated states are metastable states, because the coulombic attraction between the positive charge and the electron eventually leads them to combine and to form a neutral molecule. However, they can be important intermediates for the migration of electronic energy. As an illustrative example, let us consider two different mechanisms of energy migration in a condensed phase which consists of solvent molecules A and solute molecules B. A mechanism very frequently suggested involves only neutral excited states[†,32]:

(1) $$A \rightsquigarrow A^*$$

(2) $$A^* + A \rightarrow A + A^*$$

(3) $$A^* + B \rightarrow A + B^*$$

Step (1) is the initial excitation process, step (2) is the excitation migration (discussed in Section 4.4.2) and step (3) is the final localization of excitation energy on a solute molecule. In general the excitation energy must go through a succession of steps (2) before it can be localized; it follows that the lifetime of A^* must be sufficiently long to permit travel of excitation through A. It is also clear that in order for step (3) to occur the excitation energy of A^* must be greater than or equal to that of B^*.

An alternative mechanism involves charge-separated states.

(4) $$A \rightsquigarrow A^+ + e^-$$

(5) $$A^+ + A \rightarrow A + A^+$$

(6) $$A^+ + B \rightarrow A + B^+$$

(7) $$B^+ + e^- \rightarrow B^*$$

The steps (4), (5) and (6) are analogous to the steps (1), (2) and (3) in the previous mechanism. For such a mechanism to be probable in a two-component system, the positive ion should be the unconverted (parent) one; if the positive ion A^+ is unstable, the propagation chain represented by step (5) does not last very long. It is also necessary that the charge-recombination process be slow in comparison to the propagation step; otherwise, the charge recombination between A^+ and e^- will merely precede the excitation transfer mechanisms (1), (2), (3). Implicit in the notion of such a chain is the idea that the electron released in step (4) is trapped by some process in a region remote from the original A molecule

† Cf. the case of a scintillator dissolved in benzene[32].

(e.g. in a molecular vacancy or void)[33,34]. If the electrons can be trapped temporarily at such sites, the charge recombination can be very slow.

In the example cited, it is assumed that the positive ion is the charge carrier from solvent to solute molecule. This mechanism is, in a sense, the mirror image of the situation which can ensue if the negative charge is trapped to form a stable, movable negative ion and the positive ion is trapped in a fixed position.

An interesting case of the consequences of charge 'diffusion' is afforded by the system c-stilbene dissolved in cyclohexane. In this case, the high-energy-radiation sensitized isomerization is explicable in terms of an intermediate charge-transfer process (a chain process) in which an exo-thermal succession of isomerizations occurs[35]. (Information obtained since this writing may contravene this precise mechanism—Authors.)

4.4.2 Excitation Migration

In this entire universe we cannot establish the existence of a truly stationary state. All molecules are coupled and the degree of that coupling determines how non-stationary a state may be. In a system of exactly like molecules the coupling can be very strong and the localized state would, in conse-quence, be very non-stationary; i.e. the 'stationary state' involves all the molecules of the system. In that case (of strong coupling of molecules) energy of the excited state travels freely among them; i.e. excitation migration occurs as a radiationless transition.

There are two extreme mechanisms for excitation migration just as for charge transport in condensed media: a hopping mechanism and an exciton-band mechanism. In the hopping mechanism, the excitation hops from molecule to molecule in a 'random-walk' fashion and the range covered by the excitation migration in a time period t is given by a sphere whose radius is $\sqrt{(6Dt)}$ where D is the 'diffusion' constant. On the other hand, the case of the exciton-band mechanism, excitation migration takes place in the form of wave-packet spreading[36]; the spread of a wave-packet in a time period t is given by $2\beta t/\hbar$, where β, here, is the transfer integral introduced later in this discussion (see Equation 4.23).

In order to see the essential features of these two mechanisms, let us consider a pair of identical molecules M_1 and M_2, between which excita-tion transfer can take place. The Hamiltonian for the system can be written as

$$H = H_1 + H_2 + H_{12} \qquad (4.22)$$

where H_1 and H_2 are the Hamiltonians for the isolated molecules M_1 and M_2, and H_{12} is the interaction between them.

The probability of excitation transfer per unit time must be calculated by two different schemes, depending upon the relative magnitude of the transfer integral

$$\beta \equiv \int \phi_{M_1}\phi'_{M_2}H_{12}\phi'_{M_1}\phi_{M_2}\, d\tau_e \qquad (4.23)$$

where the primed and unprimed functions denote the electronic wave functions for the excited and ground states, respectively.

If the relaxation (transfer) time for the electronic excitation, h/β, is longer than the lifetime of the vibronic level of the individual molecule, given by $h/\Delta\varepsilon$[37] (see Figure 4.6 for definition of $\Delta\varepsilon$), the probability for

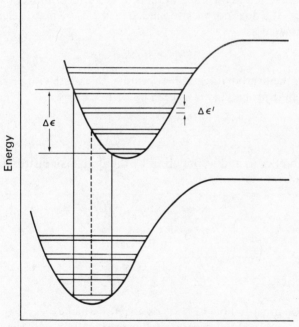

Nuclear coordinate

Figure 4.6 Potential curves for an isolated molecule with shift in equilibrium separation of excited state. The quantities $\Delta\varepsilon$ and $\Delta\varepsilon'$, introduced by Förster, are illustrated

excitation transfer can be calculated by the well-known formula of time-dependent perturbation theory[38]

$$P = \frac{2\pi}{\hbar} \frac{1}{\Delta E} \sum_i \sum_j |H'_{ij}|^2 w_i w_j \qquad (4.24)$$

where ΔE is the energy width at the final state, w_i, w_j are the probabilities that the ith and jth vibrational levels are initially occupied and the matrix element H'_{ij} is given by

$$H'_{ij} = \int \beta(\eta_i^* \eta'_j)^2 \, d\tau_n \qquad (4.25)$$

where η_i, η'_j are the vibrational wave functions specified by the quantum numbers i and j for the ground (unprimed) and excited (primed) electronic state, respectively, and the integration is to be performed over the nuclear coordinates. If β does not vary too much in (4.25), the matrix element can be approximated as

$$H'_{ij} \approx \beta \cdot S_{ij}^2 \qquad (4.26)$$

where S_{ij} is the overlap integral between the ith and jth vibrational levels. In this case the probability (4.24) can be written as

$$P = \frac{2\pi}{\hbar} \frac{\beta^2}{\Delta E} \sum_i \sum_j S_{ij}^4 w_i w_j \qquad (4.27)$$

The probabilities w_i and w_j are given by the Boltzman distribution functions. Because

$$\sum_i S_{ij}^2 = \sum_j S_{ij}^2 = 1 \qquad (4.28)$$

and

$$\sum_i w_i = \sum_j w_j = 1 \qquad (4.29)$$

the double summation of (4.27) yields a factor which is less than unity.

Kubo[38] developed the generating-function method which is useful in calculating the double summation of (4.24) and thus in studying the temperature dependence of the transition probability. In order to write the result of such calculation in simple closed form, one must introduce further approximations. Inasmuch as direct theoretical calculations of

(4.24) are not possible because of the lack of accurate molecular wave functions, it is more significant to consider the physical meaning of the double summation. According to Förster[37], the double summation represents the mutual overlap between the fluorescence and absorption spectra of the molecule. It follows without a detailed computation that the transition probability (4.24) increases with temperature and one can consider the activation energy for excitation migration in a phenomenological way (strictly speaking, such an activation energy is not temperature independent).

So far we have been considering the cases which are characterized by $\beta \ll \Delta\varepsilon$ (Figure 4.6) and are classified as weak interaction by Förster[37]. In this mechanism for excitation migration in a condensed phase, the excitation becomes localized on a molecule before it moves to another one and does not remember which molecule it came from. These are the typical characteristics of any diffusion-like motion.

On the other hand, if the relaxation time for the electronic excitation is much shorter than the lifetime of the excited state of the individual molecule ($\beta \gg \Delta\varepsilon$), Equation (4.24) is not useful for discussion of excitation transfer processes. The main reason for this situation is that the *zeroth* order electronic wave functions must now be the eigenfunctions of the electronic Hamiltonian which include the interaction H_{12}. Such eigenfunctions can be written to first order as:

$$\phi_S = \frac{1}{\sqrt{2}}(\phi_{M_1}\phi'_{M_2} + \phi'_{M_1}\phi_{M_2})$$

$$\phi_A = \frac{1}{\sqrt{2}}(\phi_{M_1}\phi'_{M_2} - \phi'_{M_1}\phi_{M_2}) \qquad (4.30)$$

It is possible to have a radiationless transition between ϕ_S and ϕ_A of (4.30); the transition probability is given by the formula

$$P = \frac{2\pi}{\hbar}\rho|H'_{SA}|^2 \qquad (4.31)$$

which is the generalized form of (4.24). In (4.31), ρ is the density of states for the final state (say, ϕ_A) and H'_{SA} is the matrix element between ϕ_S and ϕ_A with the Born–Oppenheimer operator which contains the first and second derivatives of the wave functions with respect to the nuclear coordinates. It is clear from the definition of ϕ_S and ϕ_A that such a radiationless transition does not correspond to an excitation transfer: the probabilities of finding the excitation either on M_1 or on M_2 are equal for both the initial and the final states.

In order to consider the excitation transfer in the case of strong interaction ($\beta \gg \Delta\varepsilon$), there must be a mechanism for producing a state in which the excitation is localized on one molecule initially. We do not consider such a mechanism in detail, but excitation by high-energy particles is more likely to produce such non-stationary states than is excitation by light in photochemistry.

We consider the limiting case of the strong interaction, $\beta \gg \Delta\varepsilon$, where the electronic interaction β is almost a constant with respect to the nuclear coordinate so that the nuclear wave functions associated with the symmetric and anti-symmetric electronic states, ϕ_S and ϕ_A of (4.30) have the same form. In this case, the nuclear wave functions can be dropped from the formulation and the time-dependent wave function $\Psi(t)$ can be expressed as

$$\Psi(t) = C_S \phi_S \exp\left(-\frac{iE_S t}{\hbar}\right) + C_A \phi_A \exp\left(-\frac{iE_A t}{\hbar}\right) \tag{4.32}$$

where C_S and C_A are time-independent coefficients which can be determined from the initial condition, and E_S and E_A are the electronic eigenvalues associated with ϕ_S and ϕ_A. If the excitation is initially localized on M_2,

$$\Psi(0) = \phi_{M_1} \phi'_{M_2} \tag{4.33}$$

Then the coefficients C_S and C_A must be equal to $1/\sqrt{2}$. From (4.30) and (4.32)

$$\Psi(t) = \tfrac{1}{2}\left\{\phi_{M_1}\phi'_{M_2}\left[\exp\left(-\frac{iE_S t}{\hbar}\right) + \exp\left(-\frac{iE_A t}{\hbar}\right)\right]\right.$$

$$\left. + \phi'_{M_1}\phi_{M_2}\left[\exp\left(-\frac{iE_S t}{\hbar}\right) + \exp\left(-\frac{iE_A t}{\hbar}\right)\right]\right\} \tag{4.34}$$

The square of the coefficient of $\phi'_{M_1}\phi_{M_2}$ gives the probability of finding the excitation of M_1 at time t; thus,

$$P(t) = \sin^2\left(\frac{E_S - E_A}{\hbar} \cdot t\right) \tag{4.35}$$

From

$$E_S - E_A = 2\beta \tag{4.36}$$

$$P(t) = \sin^2\left(\frac{2\beta t}{\hbar}\right) \tag{4.37}$$

It is clear that the excitation transfer is oscillatory with the period $4\beta/h$.

When the system consists of a large number of identical molecules in a linear array with strong interaction, the probability of finding the excitation on a given molecule at time t is again oscillatory, but is given by a Bessel function[36]

$$P(t) = \left[J_n\left(\frac{2\beta t}{\hbar}\right)\right]^2 \tag{4.38}$$

where n is the number of the primitive translations between the initially excited molecule and the molecule under consideration. Excitation migration takes place, in this case, as coherent spreading of an exciton wave-packet. In this mechanism of energy transfer, any factor which tends to destroy the coherence of the exciton wave will decrease the rate of excitation migration; such factors include molecular vibrations and irregularities in the lattice. As a very important example, increase of temperature specifically decreases the rate of excitation migration.

A more detailed, and therefore realistic, mechanism of energy transport would include the effect of molecular vibrations. Even if the system responds initially as one of strong interaction, intra- and intermolecular vibrations will eventually change it to one of weak interaction; excitation or charge will eventually be self-trapped. Such a transition from strong to weak interaction in the mechanism of excitation migration will require a time of the order of the relaxation time, which is rather short for most liquids and crystals (10^{-12} to 10^{-10} s), but can be quite long, 30 min or more for viscous liquids or glasses at low temperatures.

The weak and strong couplings represent two extreme cases of energy transfer. The majority of molecular aggregates of chemical and biological interest probably falls into the range of weak to intermediate coupling.

4.4.3 Charge Migration

The subject of charge migration in condensed systems is so vast that it must be beyond the scope of any limited description such as this. However, there are certain limited analogies to which attention can be properly and profitably addressed. It is worthwhile, as an example, to note the exact analogy between the modes of migration of excitations and of *isolated* charges, whether they be electrons or holes.

The essential mechanism for charge migration between two sites is provided by the overlap of charge distribution functions (or wave functions) between them. Dependent upon the size of this overlap, two extreme mechanisms of charge migration must be considered.

The hopping mechanism corresponds to the weak-coupling case for excitation migration. The rate of charge migration in such a case is so small that the surrounding medium has time to relax completely each time the charge migrates from one site to another. The range covered by the charge in time t is given by a sphere of radius $\sqrt{(6Dt)}$ where D is the appropriate diffusion constant for the random walk. Increase of temperature increases the rate of charge migration as does any other activation process.

The other extreme case is the band mechanism. In this case the rate of charge migration is so fast that relaxation of the medium cannot follow the motion of charge. Increase of temperature in such case decreases the rate of charge migration.

A more complete statement of mechanism requires explicit consideration of the effect of molecular vibrations on electronic motions*. It should also be noted that the isolated charge does not exist under the conditions of the radiation chemistry of condensed systems (see Section 4.2). It is not even proper to ignore electron–electron correlation at low dosage except for specific isolated portions of the irradiated system. Further, dynamical correlation of positive charge and electron cannot be ignored under any circumstance.

4.4.4 General Comment on Electronic Energy Transfer

At present, there is a fair understanding of the electronic states which may be produced in the radiation chemistry of condensed systems (cf. Sections 4.2 and 4.3), their distribution (cf. Section 4.2) and of the time and possible duration of their existence (cf. Section 4.3). Further, it is shown in this section, the theory of the behavior of excited states and of isolated charged species under limited (but extreme) conditions is in the course of rather sophisticated development. The difficulty is that the models employed

---

* It may be mentioned that this is one of the important, central problems of the solid state still commanding the attention of numerous theorists.

for interpretation of the various processes which can be imagined to occur do not conform too closely to reality. Thus, in spite of the fact that allusion may be repeatedly made to well-developed theory, the fact is that the statements made are, perhaps necessarily, frequently in the area of speculation. In the original meaning of the term they may be 'sophisticated' and therefore impressive. However, for actual interpretation of experimental facts the supporting theory must be based on somewhat less detailed, more superficial models which do not put too much demand on actual knowledge of microscopic details.

4.5 ENERGY TRANSFER TO SCINTILLATORS

Perrin and Choucroun[39] were the first to show luminescence involving excitation transfer (between dye molecules) in liquid systems. Interest in such transfer processes in high-energy-excited liquid systems stems from the work of Kallmann and Furst[40]. Data bearing on the specific rates of the various processes involved derived initially from the ultra-fast luminescence decay time studies of Burton and coworkers[41,42,43,44]. The theoretical foundations of resonant energy transfer were established in the early work by Perrin and Choucroun[39] and by Kallmann and London[45] for sharp-line systems and for rigid systems by Förster[46]. The theory of energy transfer was extended by Dexter[47] and again by Förster[48]. The applicability of Förster's theory to liquid systems has been discussed by Salivanenko[49] and by Povinelli[50].

4.5.1 Real Cases

Section 4.4 examined energy transfer processes in the special cases of strong and weak coupling. In practical application of such results to luminescent systems, strong coupling may be observed in crystals and in those liquid solutions of dyes the absorption spectra of which change significantly with concentration[46]. In radiation chemistry, the existent experimental work on liquid mixtures has been confined to cases where weak-coupling considerations govern. Similarly, the luminescence studies of significance in the radiation chemistry of liquids are also in this category. This section is consequently limited to examination and interpretation of the observable luminescence phenomena which may be associated with energy transfer in weakly-coupled systems either of excited solvent and solute or of two solutes one of which is initially excited.

In solutions generally, we are concerned further with diffusing systems in which energy may be transferred in actual collision processes, the kinetics of which have been exhaustively studied. The pertinence of such results to the luminescence problem at hand is examined. On the other hand, the more interesting question of the actual mechanism of excitation transfer or quenching in such cases is not examined in detail.

A third case of excitation transfer in condensed systems involves the actual emission of light by one molecule of the system and reabsorption by another. Although it can be experimentally important, it is theoretically trivial and is actually called 'the trivial process' by workers in this field. It is examined only lightly in this section.

In general, as was shown in Section 4.2, high-energy radiation excites a great variety of states. Nevertheless, it is a curious fact of luminescence processes resultant directly or indirectly from such excitation that with a single exception (namely azulene) the luminescence from complex molecules appears to be the accompaniment of transition from the lowest excited singlet state to the ground state. The preferred explanation involves a high probability of internal conversion from higher excited singlet states to the first[51]; in the case of the single exception, the large energy difference between the second and first excited states seems to preclude such transition in the period before emission involving transition from the second excited state to the ground state[52].

It seems to be established that chemical processes may involve higher states and even ionized states of initially excited molecules (cf. Section 4.6). Thus, it may be expected that some luminescence processes can ensue from energy transfers involving higher excited states or ionized states. Conditions under which such processes can occur must, therefore, also be examined.

In many cases, if a solvent absorbing radiant energy contains small amounts of fluorescent additives, luminescence characteristic of such additives can be observed with a yield exceeding that expected on the basis of direct excitation of the additive and its known quantum efficiency for fluorescence. Energy, therefore, must have been transferred from the matrix to the additive. The transfer process can result either from absorption of photons emitted by the matrix if the latter luminesces (i.e. the trivial case), from collisional interaction between additive and matrix, or from resonance interactions.

Such transfer processes are not characteristic of high-energy-excited systems alone. They are possible in the case of excitation by non-ionizing radiation. In the latter case the resulting luminescence (i.e. fluorescence) is more easily accessible to experimental and theoretical investigations. Interpretation of energy-transfer processes, therefore, must be consistent with the results obtained under such conditions.

Characteristics peculiar to high-energy-excited systems may be expected whenever an ionic reaction is involved in the luminescence process and whenever the possible effect of non-homogeneous distribution of reactive species (cf. Section 4.2) must be considered. In this section the excited species from which energy is transferred is called the donor while that which receives the energy is denoted as the acceptor. Unless specifically indicated it is assumed that donor and acceptor concentrations are small (of the order of $10^{-3} M$ for acceptor and probably much less for donor). The reference, of course, is to average concentration of the donor species. Under conditions of high-energy irradiation it is possible that the local concentration of a donor species D is of the same order of magnitude as the species M from which it is formed by the process $M \rightsquigarrow D$ (cf. spurs, etc., Section 4.2).

The mechanisms of energy transfer here summarized are : energy transfer by emission and absorption of photons, energy transfer in collision, resonance transfer processes, and ionic transfer processes. While the first three processes are common for systems excited both by non-ionizing and by ionizing radiation, the fourth can play an important role only in the luminescence of high-energy-excited systems. Although the importance of ionic energy transfer processes is experimentally well established, a quantitative description is still incomplete (cf. Section 4.4.3).

The detailed discussion of energy transfer processes involving 'initial excitation'* is based on observations on excitation by non-ionizing radiation. Direct applicability to high-energy excited systems must be separately established in each case. Nevertheless, it seems that the overall luminescence behaviour of such systems is very often similar to that excited by non-ionizing radiation.

4.5.2   Energy Transfer Processes in Scintillator Systems

Radiative energy transfer ('the trivial process')

Fluorescent molecules can interact with each other through their common radiation fields. If the absorption spectrum of the acceptor overlaps the emission spectrum of the donor an emitted photon may be reabsorbed.

---

* Note that in this sense the term 'initial' is loosely defined; cf. Section 4.3.

Although this process is not normally considered in transfer processes it must nonetheless be taken into account in experimental work. Because the degree of such reabsorption depends on the geometry of the system it can be distinguished from other transfer processes by variation of the experimental set-up. For accurate determination of fluorescence spectra, luminescence efficiencies or luminescence decay, the effects of radiative energy transfer must be excluded.

Energy Transfer in Collision

In liquid systems, where relatively rapid diffusion is possible, energy transfer can occur on collision of donor and acceptor molecules. The mathematical formalism to describe the rate of the transfer process is then analogous to that of bimolecular reactions in condensed phases. Experimentally this situation is frequently encountered in fluorescence quenching when donor and acceptor do not possess energy levels suitable for resonance transfer (e.g. in the case of excited p-terphenyl quenched by carbon tetrachloride).

The simplest theoretical treatment[53] of such bimolecular reactions leads to an expression for the specific rate k of this process

$$k = 4\pi R D N' \tag{4.39}$$

in units of litre $\text{mole}^{-1}\,\text{s}^{-1}$ where R is the radius of interaction, D the mutual diffusion coefficient of donor D and acceptor A (i.e., $D = D_D + D_A$) and N' is Avogadro's number divided by 1000.

Consider the situation around a single molecule of *unexcited* donor species M. It is randomly surrounded by other molecules M and acceptor molecules A. The probability of finding a molecule of A at distance r from a molecule M is the same throughout the solution. When a single molecule M is excited

$$M \rightarrow D$$

and the probability of finding a molecule A at distance r from D becomes a function of time t. Specifically, the situation at initial time t_0 depends very much on the probability of the reaction $D + A \rightarrow M + ?$ when the species D and A actually collide. On the assumption that that probability is unity*, it follows that at t_0 (under any circumstance) any molecule D in contact with A disappears. Thus, at the very least, this portion of the distribution is missing. Further, if the time required for collision of D and

---

* i.e. the basic assumption of the Smoluchowski equation.

A be sufficiently short, there is a complete absence of experimentally detectable species A within a distance r_0 of D.

Equation (4.39) is derived under the assumption that the initial distribution of donor and acceptor is instantaneously reestablished after each transfer process; i.e. diffusion is very fast. This condition is in general true in gas phase reactions and very nearly so in liquids of very low viscosity.

A more general treatment must take into account the depletion of excited molecules in the immediate neighborhood of a quencher molecule. The establishment of a stationary distribution of reactants, therefore, requires some time. Taking this effect into account, Förster[54] has given for the specific rate

$$k = 4\pi N'RD[1 + R/(Dt)^{1/2}] \qquad (4.40)$$

The second term in (4.40) describes the initial transfer process. Assuming the lifetime of the excited donor to be 10^{-8} s, $R = 10$ Å and $D \sim 10^{-5}$ cm s^{-2}, this initial part contributes about 30% to the total quenching process. In the case of viscous media and for short lifetimes this effect is, therefore, not negligible. A detailed discussion of this problem has been given by Sveshnikov[55].

Experimentally the rate constants are determined from measurement of luminescence intensity under steady-state conditions and from luminescence decay curves as functions of the various ambient conditions. In cases where initial transfer is negligible (e.g. when R is very small and D is large), luminescence intensity I, decay time τ and luminescence quantum yield γ are related to the acceptor concentration C_A by the Stern–Volmer relation:

$$I/I_0 = \gamma/\gamma_0 = \tau/\tau_0 = 1/(1 + k\tau t_0 C_A) \qquad (4.41)$$

where the subscripts refer to the corresponding quantities in the absence of acceptor.

In cases where initial transfer is important (cf. Equation 4.40) a simple relation such as (4.41) no longer holds. For excitation with very short pulses the decay is no longer expected to be exponential: I as a function of t then follows a decay law of the form

$$I(t) = I(0)\exp\{-t/\tau_0 - 4\pi N'DRC_A t[1 + 2R/(\pi Dt)^{1/2}]\} \qquad (4.42)$$

The result is that the luminescence decay is initially faster and at later times becomes exponential. When instead of short-pulse excitation the excitation is so prolonged that a steady state is attained, the decay appears to be experimentally logarithmic. The pertinent decay equation derived by Yguerabide, Dillon and Burton[56] is not simple but its plot does give a curve which cannot be experimentally differentiated from true exponential decay.

Resonance Transfer

The transfer mechanism discussed in the previous paragraphs requires close proximity of donor and acceptor. By contrast, the processes here considered can occur over distances of many molecular dimensions. The reabsorption process (i.e. the trivial process) involves the coupling of donor and acceptor through actual radiation emission and absorption. On the other hand, the coupling required for resonance transfer consists of mutual interaction between donor and acceptor.

Although the theories developed and sketched here apply strictly only to systems in the solid state, their important features may be retained in discussion of resonance transfer processes in the liquid phase. The system considered may be thought of as an inert matrix containing low concentrations of donor and acceptor molecules in a random distribution. The transfer process (in these weak coupling cases) may be broken up into a sequence of five steps: absorption of energy E by donor, vibrational relaxation to a state of energy $E' < E$, transfer of energy E' to acceptor, relaxation of acceptor to state of energy $E'' < E'$ (as well as relaxation of donor, now M, to an approximately unexcited state at about this same time), and possible emission of E'' as luminescence or non-radiative degradation.

Of particular interest is the transfer of energy E' to the acceptor. Vibrational *relaxation* steps generally require about 10^{-13} s. By contrast, emission generally requires about 10^{-9} to 10^{-8} s. The rate of energy *transfer* depends primarily on the donor–acceptor distances.

It is also to be noted that, in general, the reverse transfer process (i.e. from excited acceptor to deexcited donor) need not be considered here because the acceptor and donor are not in resonance after vibrational relaxation of the acceptor.

The specific rate of the transfer process can be evaluated from time-dependent perturbation theory of quantum mechanics. The general expression for the probability of transition of the initial state (excited donor D plus ground state of acceptor molecule) to the final state (ground state of the donor M and excited acceptor molecule) can be given (cf. Equation

4.31) by

$$P_{DA} = \frac{2\pi}{\hbar} \rho_E \left(\int \psi_i^* H' \psi_f \, d\tau \right)^2 \tag{4.43}$$

where ρ_E is the density of accessible states, ψ_i and ψ_f correspond respectively to the wave functions of the initial and final states of the donor-acceptor pair and both states have the same energy and H' represents the perturbation Hamiltonian.

The sum of all coulombic interactions between outer electrons and cores of donor and acceptor is taken in setting up H'. Development of H' in the form of a series in R (the distance between donor and acceptor), describes the interaction energy as a sum of dipole–dipole, dipole–quadrupole and higher-order terms[47].

Because of their small contributions, higher-order terms are usually neglected and only dipole–dipole interactions are considered. A detailed theory of dipole–dipole interaction was given by Förster[48]. When a dipole transition is forbidden, higher-order transitions may become of importance. The case of dipole–quadrupole transitions in particular has been extensively treated by Dexter[47].

Dipole–dipole interaction. The term in the expression of H' corresponding to dipole–dipole interaction has the form

$$H'(R) = \frac{e^2}{\kappa R^2} \left[r_D \cdot r_A - \frac{3(\mathbf{r}_D \cdot \mathbf{R})(\mathbf{r}_A \cdot \mathbf{R})}{R^2} \right] \tag{4.44}$$

where $\mathbf{r}_D$ and $\mathbf{r}_A$ are position vectors referring to the electrons on donor and acceptor, respectively, and κ is the dielectric constant of the medium.

If the wave functions describing the system were known, combination of Equations (4.43) and (4.44) would allow calculation of the transfer probability. In view of the complexity of the system, however, such treatment is not feasible. On the other hand, the matrix elements in question are closely related to the oscillator strength of the donor, the absorption coefficient of the acceptor and its decay time. Because these are measurable quantities in the case of allowed dipole transitions their relationship to the unknown matrix elements can be evaluated; the transition probability can thus be expressed as

$$P = \frac{9\chi^2 c^4 \ln 10}{128\pi^5 \eta^4 N' \tau_0 R^6} \int_0^\infty f_D(\nu) \varepsilon_A(\nu) \frac{d\nu}{\nu^4} \tag{4.45}$$

where η is the refraction index of medium, c is the velocity of light, τ_0 is the natural lifetime of the donor (i.e. in absence of acceptor), v is the frequency, $f_D(v)$ is the spectral distribution function (this function gives the fraction of emitted quanta within unit interval of range at frequency v), $\varepsilon_A(v)$ is the absorption coefficient of acceptor as a function of v, and χ, the orientation factor, is of the order of unity. Equation (4.45) can be written in the form

$$P = \frac{1}{\tau}\left(\frac{R_0}{R}\right)^6$$

(4.46)

with

$$R_0^6 = \frac{9\chi^2 c^4 \tau \ln 10}{128\pi^5 \eta^4 N' \tau_0} \int_0^\infty f_D(v)\varepsilon_A(v)\frac{dv}{v^4}$$

It follows from (4.46) that R_0 can be identified with a critical distance for which the probability of energy transfer is equal to that of deactivation of the excited donor by all other processes (e.g. emission, internal conversion).

The critical transfer distance is related to the experimentally important critical concentration, C_0, by

$$C_0 = \frac{3}{4\pi N' R_0^3}$$

Some typical values for critical transfer distances of commonly used scintillator systems are given in Table 4.1.

Higher-order interactions. For the case where dipole transitions are forbidden, higher-order interactions may have to be considered as involved in energy transfer. However, such interactions are effective only over much shorter distances and are, therefore, important only in pure or highly concentrated donor–acceptor systems.

Dexter[47] has extended the theory of resonance interaction to the dipole–quadrupole case. He has shown that the transfer probability is again proportional to the overlap of emission and absorption spectra of donor and acceptor, respectively, and varies with the inverse eighth power of distance between them. Unlike the dipole–dipole case, the theoretical expression cannot be related to experimental data for the simple reason that such data are not available (quadrupole transitions are less probable and the corresponding emission intensity weaker by a factor of about 10^{-7} compared to a dipole transition of the same wave length). Nevertheless, theory shows that the ratio of the probabilities of dipole–quadrupole to dipole–dipole energy transfer is of the order of $(a/R)^2$ where a is the atomic

Table 4.1

Calculated critical transfer distances R_0 for some common liquid
scintillator systems according to Voltz and coworkers[57]

Solvent	Solute	R_0(Å)
Benzene	PPO	16·5
	pT	17·5
	DPH	14·0
p-Xylene	PPO	20·0
	pT	19·0
	DPH	17·0
Toluene	PPO	18·0
	pT	18·0
	DPH	15·0

PPO = diphenyl-2,5-oxazole
pT = p-terphenyl
DPH = diphenylhexatriene

radius and the transfer may be quite significant. It follows from the relative values of a and R that, for *small* donor–acceptor distance, the higher order process can be about $\frac{1}{10}$ as probable as the dipole–dipole transfer.

In general, comparatively short distances are required for dipole–quadrupole transfer to be important. This requirement, however, makes it difficult to separate this process from other higher-order interaction or exchange effects. A discussion of the relative strength of interactions of various orders is given by Dexter[47].

An attempt to explain the abnormally high value of energy transfer and quenching constants found in systems of liquid aromatic solvents, such as benzene or toluene, as a result of a higher-order interaction has been made by Voltz, Laustriat and Coche[58].

Ionic Transfer Processes

Section 4.5.2 preceding suggests that the luminescence behaviour of high-energy-excited scintillator systems may be expected to differ from that of the ordinary u.v. excited systems because of the possible involvement of ions in the luminescence process. Indeed, experiments by Hamill and coworkers[59] on glassy systems at low temperatures (i.e. $< 100°$K) have given clear experimental evidence of the formation of ionic species under suitable irradiation conditions and it may be expected that similar species can be formed, with much shorter half life, even in liquid systems. Unfortunately, although the presence and involvement of ionic species seems well established, the theory of energy transfer via such species is, as Section 4.4.3 indicates, inadequate for quantitative treatment of the data. Instead, the procedure of interpretation of the results is still essentially qualitative, inferential and speculative.

A rough survey of the available experimental data seems to indicate that the differences between u.v. and high-energy excited luminescence effects are frequently quite small. For example, the luminescence decay time of excited p-terphenyl in benzene is the same for both kinds of excitation. In both cases, Stern–Volmer plots relating luminescence intensity to quencher concentration seem to be valid over a wide range of the latter. A characteristic function in luminescence is defined by

$$\gamma \equiv \frac{k_q}{\sum_i k_i}$$

where k_q is the specific rate of quenching and $\sum_i k_i$ is the sum of specific rates of all other decay processes (expressed in suitable units); the experimental value of this function is very closely the same for both kinds of excitation. The similarity of these results seems to suggest that, if ionic reactions are involved in luminescence processes, such reactions are so fast compared to the emission process that the overall luminescence behaviour is essentially determined by the reactions of the excited scintillator. More refined experimental techniques reveal differences in some aspects of the luminescence behaviour of liquid scintillator systems, with the degree of difference dependent on the particular system under investigation.

Figure 4.7 represents an example of the type of difference in decay characteristics shown by u.v. and high-energy-excited scintillator systems. It shows the luminescence decay of 10^{-3} M p-terphenyl in cyclohexane excited by 1 ns pulses of u.v. ($\lambda = 2537$ Å), on the one hand, and of 30 kV x-rays on the other. The initial part of the decay is the same in both cases (and is a fixed characteristic of the scintillator itself)[60]. In addition, however, the high-energy-excited system shows a long component (not found in the u.v. case); similar contrast of behaviour is shown in all (liquid) scintillator solutions which have been studied.

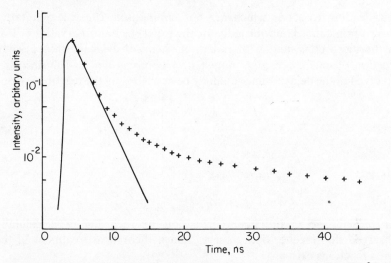

Figure 4.7 Comparison of typical luminescence decay curves for $10^{-3} M$ *p*-terphenyl in deaerated cyclohexane. ——— Fluorescence decay after excitation with ~ 1 ns pulse of 254 nm light. + + + + + + Luminescence decay after excitation with a 1 ns pulse of 30 kV x-rays

Some indication of the origin of the long decay component in luminescence can be obtained from studies of the effect of quenchers. For example, the addition of small amounts of carbon tetrachloride, which from radiation chemical studies is well known to act as a negative-ion scavenger, not only results in an overall reduction in luminescence intensity but also affects the slope and form of the long component. Similar behaviour of this component is observed when positive-ion scavengers are added to the system. From the absence of any measurable effect of these additives (at the concentrations employed) on the u.v. excited systems, it may be concluded that interaction of the additive with the excited scintillator does not occur under such conditions. A corollary conclusion is that, in high-energy-excited systems, the interaction occurs with a precursor of the excited scintillator. From the general character of the quenchers added it may be assumed that these precursors are, at least in part, scintillator ions. In the absence of scavenger their neutralization can yield excited scintillator molecules which are ultimately responsible for the delayed emission. In the presence of quencher, neutralization of scintillator ions by non-quenchers (e.g. electrons) competes with that by quencher ions leading to states which are not luminescent. Such a qualitative description can account broadly for the experimental observations. It is furthermore consistent with radiation chemical evidence that a small fraction of ions can escape immediate recombination[61]. The processes involved in the delayed emission may be described by the reaction scheme

$$(1) \qquad X^+ + X^- \rightarrow \begin{bmatrix} X + X^* \\ \text{or} \\ X_2^* \end{bmatrix} \rightarrow 2X + h\nu$$

$$(2) \qquad X^+ + Q^- \rightarrow X + Q$$

$$(3) \qquad X^- + Q^+ \rightarrow X + Q$$

At present, no quantitative correlation between the experimentally observed fluorescence decay curves and a theoretical treatment of the ionic reactions exists.

It has recently been noted (in some unpublished work by Huque and Ludwig) that the long component of the luminescence of the scintillator system in the absence of quenchers can be represented by a time-intensity relationship characteristic of bimolecular reaction between ions. The difficulty in establishing a quantitative relationship arises from uncertainties in the theory of ionic reactions and from the complication that one is not dealing with an initially homogeneous distribution of reactants.

From the (decay-time) constancy of the exponentially decaying part of the luminescence irrespective of concentration of scintillator in *aliphatic* solvents, it follows that the rate determining process is the emission. The formation of the excited state must therefore be fast; in the case of cyclohexane solvent[62] it was shown to occur in less than 0·3 ns. A treatment of the formation of excited scintillator in this case as resulting from collisional energy transfer from 'excited' cyclohexane (in a purely formal sense) to the scintillator leads to rate constants of the order of 10^{11} to $10^{12} \, M^{-1} \, s^{-1}$. Such high values are not only inconsistent with what can be expected from theory of diffusion-controlled reactions but also with the fact that no sufficiently long-lived excited state of cyclohexane has yet been found to exist in the liquid state. On the other hand, with the assumption that ionic reactions are responsible for the formation of excited scintillator states (cf. reaction 1 above), the very fast formation of such states is at least qualitatively understandable. The potential energy of ions, resulting from coulombic interaction, is considerably higher than their thermal energy even at relatively large distances of separation so that, whenever two ions of opposite charge are separated within a certain critical distance (e.g. for liquids of low dielectric constant, about 100 Å) the recombination time becomes immeasurably small. In the case of high-energy radiation, where the reactants are initially localized within possibly even smaller regions (cf. Section 4.2) the separation of the reactants is correspondingly smaller and coulombic interaction is very strong. In such situations, therefore, the (ionic) reaction rates can be expected to be immeasurably large. Thus, if such ionic reactions are involved in the formation of excited scintillator states, that fact is not revealed and the initial part of the decay curve is smoothly exponential. On the other hand (as already implied in reference to unpublished work of Huque and Ludwig), delayed neutralization (the result of wide separation of charge) is a second-order process and is the reaction which determines the rate of light emission in the low-intensity luminescence tail under conditions of high-energy excitation.

4.6 ENERGY TRANSFER EFFECTS IN CHEMICAL YIELDS

The classical experiment of Franck and Cario[63] showed that, in the vapour phase at least, an excited species (in that case, the 6^3P_1 Hg atom) can transfer energy to a (hydrogen) molecule, with resultant chemical effect. Later, West and Paul[64] showed that the quantum yield of photo-decomposition products of alkyl iodides in benzene solution can be explained in terms of excitation transfer from excited solvent to iodide. The first indication of the existence of a similar phenomenon in radiation chemistry derives from the finding by Schoepfle and Fellows[65] that the yield of hydrogen in electron-bombarded decomposition of cyclohexane is decreased in benzene solution to a level below that which might be expected from a simple mixture law. In consideration of that result and on the basis of somewhat more extensive data of their own, Manion and Burton[66] proposed an explanation based on the transfer of energy (as ionic or electronic excitation) from the cyclohexane to the benzene.

In general, arguments for energy-transfer effects in chemical yields derive from three somewhat different, but related, kinds of investigation (other than luminescence studies, cf. Section 4.5): (1) studies of the variation in yield of some product or products characteristic of one component of a binary mixture over the entire range of mixture composition; (2) sensitization studies in which the yield of a product characteristic of a

solute is measured at relatively low concentrations of the solute ($<0.5\ M$);
(3) pulse radiolysis studies in which yield of a transient species formed
from a solute at relatively low concentration is measured as a function of
time, usually by absorption spectrophotometry. The mechanism of such
energy-transfer processes may involve any or all of three kinds of elementary process: (1) collective excitation in condensed phases with subsequent
preferential localization of energy on a particular component of the
system[67] (cf. Sections 4.3 and 4.4); (2) excitation transfer from neutral
excited states of individual molecules (cf. Section 4.4); (3) transfer of
energy or reactivity via positive-charge transfer or electron-capture
processes (cf. Section 4.4). In this section, attention is focused on selected
experimental studies which provide examples of the kinds of evidence,
or arguments, presented for the role of energy transfer in radiation
chemistry.

4.6.1   Evidence for Processes Requiring Energy Transfer

It must be emphasized that excitation or ionization transfer in radiation
chemistry usually has a very speculative basis. First, there must be some
experimental indication that no 'ordinary' chemical phenomenon (e.g. a
free-radical reaction or series of reactions) can account for the full array
of kinetic features. Second, the process suggested must be physically
possible (e.g. an ionization transfer in the direction of decreasing ionization
potential). Third, if primary localization from initially spread-out
excitation is suggested, some supporting experimental evidence for such
a view is desirable. The simple fact is that the last is the most difficult to
demonstrate; its basis is in fundamental physical theory (cf. Section 4.4)
but such theory has received strong support[68].

The original suggestion of ionization transfer to explain protection in
cyclohexane-benzene systems[*,66,23] was based on the assumption that
initial deposition of energy in a single component of a low-atomic-weight
system of substances is determined by the electron fraction† of that
component and that preferential initial deposition cannot occur‡. Burton,
Hamill and Magee[69] suggested that fractional molecular polarizability
would be a better choice than electron fraction and this view has been

---

* Possible excitation transfer in that case has been ruled out on the basis that excited
cyclohexane has too short a half-life[23].

†The electron fraction ε of a component A in a mixture is defined as the fraction of the
total number of electrons (in the mixture) associated with that component.

‡ Numerous arguments against this assumption have been offered and alternative proposals have been made[69].

supported by Klots[70]. Burton and coworkers[71] offered telling arguments that no explanation involving such simple chemical effects as H atom trapping by the benzene or even intermediate trapping can explain the effects. The possibility of such intermediate reaction was ruled out by the work of Dyne and coworkers[72], who showed also that (at least) two processes of different order are involved in the primary decomposition of cyclohexane and that they are inhibited by benzene to a comparable extent.

In brief, a rather sophisticated argument must be presented in each case in which excitation or ionization transfer is invoked to explain either sensitization or protection. It is not the intention of this chapter to review all those arguments. Instead, Table 4.2 lists certain illustrative cases for which such suggestions have been made as well as the general types of evidence offered.

It should be emphasized that spectrometric data have been offered in evidence of the existence of intermediates, that excited states involved can include ion pairs, singlets and triplets, and that the special consequences of the geometry of energy deposition (spurs, blobs, etc.; cf. Section 4.2) in radiation chemistry play an important role.

4.6.2 Extended Energy Transfer

In the cases suggested in Section 4.6.1, the energy transfer occurs in each case in a single step and the process is thereafter terminated. The evidence of transfer becomes much more compelling when the effects cannot possibly be explained on the basis of speculative proposals regarding preferential deposition of energy[69] or preferential localization after the initial deposition. Such a case is exemplified by radiation-induced $cis \rightarrow trans$ isomerization of stilbene[80].

In the radiation-induced $cis \rightarrow trans$ isomerization of stilbene in either benzene or cyclohexane, evidence is provided for a process that may be broadly interpreted as a kind of energy-transfer (or reactivity-transfer) chain. Values of $G_{c \rightarrow t}$ can become quite large at c-stilbene concentrations of $\sim 0.1\ M$ and greater. For example, with very pure reagents and at low doses $(3 \times 10^{18}\ \text{eV ml}^{-1})$, $G_{c \rightarrow t} = 210$ for $0.6\ M$ c-stilbene in benzene[91] and $G_{c \rightarrow t} = 600$ for $0.2\ M$ c-stilbene in cyclohexane[92]. Such values cannot possibly be interpreted in terms of a non-chain process involving any single species producible by irradiation. The evidence implicates an ionization-transfer chain initiated by the reaction in the solvent S

$$S \rightsquigarrow S^+ + e^-$$

Table 4.2

Some examples of energy transfer in the radiation chemistry of liquids

System	Type of transfer	Evidence[a]
I. Binary Mixtures		
Six mixtures of saturated organic compounds[73]	Positive charge	Deviations of $G(H_2)$ from a 'law of averages'
(a) Benzene with *3* cycloalkanes, (b) benzene with *2* arylcyclo-alkanes, and (c) *2* cycloalkanes with their aryl derivatives[74]	Unspecified	Suppression of $G(H_2)$
Dioxane-1,4 and benzene[75]	Unspecified	Suppression of $G(H_2)$ and G(polymer)
Thiophenol and benzene-d_6[76]	Positive charge and excitation	Enhancement of $G(H_2)$ and $G(C_6H_5S_2C_6H_5)$
Nickel tetracarbonyl and benzene[77]	Positive charge or excitation transfer or primary localization	Behaviour of $G(H_2)$ and $G(C_6H_5C_6H_5)$
II. Dilute Solutions		
HI in cyclohexane-d_{12}[78]	Electron capture	Formation of H_2
N_2O in cyclohexane[79]	Electron capture	Formation of N_2
t-Stilbene in benzene[80]	Excitation (largely triplet)	Sensitization of isomerization
t-Stilbene in cyclohexane[80]	Electron capture and positive charge	Sensitization of isomerization
Butene-*2* in benzene[81]	Triplet	Sensitization of isomerization
ND_3 or C_2H_5OD in cyclohexane[82]	Proton	Formation of HD
CCl_4 in cyclohexane[83]	Electron capture	Formation of Cl^- and suppression of $G(H_2)$
$FeCl_3$ in benzene[84]	Excitation	Reduction of $FeCl_3$
Metal perphenyls in benzene[85]	Excitation	Enhancement of $G(C_6H_5C_6H_5)$
III. Pulse Radiolysis of Dilute Solutions		
Naphthalene, anthracene, or 1,2-benzanthracene in acetone[86]	Triplet	Kinetics of growth of solute triplet absorption spectrum
Anthracene and dimethyl fumarate in benzene[87]	Triplet	Observations of anthracene triplet absorption spectrum

Table 4.2 (cont.)

System	Type of transfer	Evidence[a]
Nine different aromatic solutes in cyclohexane[88]	Electron capture and positive charge	Spectra of solute radical ions
Naphthalene or benzophenone in benzene and cyclohexane[89]	Excitation and electron exchange	Spectra of solute triplets and free radicals
Variety of solute–solvent systems[90]	Excitation and electron capture	Spectra of solute triplets, solute radical anions, and free radicals

[a] G is defined as the yield of molecules of a particular species (indicated in parentheses) decomposed or produced per 100 eV input.

and propagated by the stilbene anion or cation or both as follows:

$$S^+ + c\text{-St} \rightarrow S + t\text{-St}^+$$

$$e^- + c\text{-St} \rightarrow t\text{-St}^-$$

$$t\text{-St}^+ + c\text{-St} \rightarrow t\text{-St} + t\text{-St}^+$$

$$t\text{-St}^- + c\text{-St} \rightarrow t\text{-St} + t\text{-St}^-$$

Termination is by ion-pair neutralization or charge transfer to an impurity or both, depending on dose rate and impurity concentrations. With the plausible assumption that 'free' charged species (formed with $G = 0.1$) are the chain initiators, chain lengths can be calculated. For the values of $G_{c \rightarrow t} = 210$ and 600, chain lengths of ~ 2000 and 6000, respectively, are obtained (cf., however, the end of Section 4.4.1).

4.7 CONCLUSION

In this chapter it is shown on a well-developed theoretical basis that high-energy radiation deposits energy in a condensed system in a non-random way (i.e. in a track) and that many of the effects characteristic of radiation chemistry are directly attributable to such non-random energy deposition. However, there is a definite time interval (approaching 10^{-14} s) between the energy deposition act and the first experimentally recognizable physical event (cf. Figure 4.5). In that interval, initial excitation which may appear to involve a single electron, may become successively delocalized (e.g. in superexcitation which may overlay several molecules or in exciton production) and then localized in a molecule or group of molecules not associated with the initial energy deposition act. However, it cannot yet be established unequivocally whether the energy localization, even for low-atomic-weight systems, is randomly (e.g. related merely to electron fraction of the components) or more specifically determined.

Even after localization of the excitation becomes experimentally meaningful, it is possible that the excitation travels as such, or as the movement of charge, to a more remote locale where the first 'chemical' effect may evidence itself in luminescence, in ionization, or in chemical decomposition or even in methathetical processes. For this stage of the processes of radiation chemistry, the theory is fairly good (in the mathematically quantitative sense) for excitation transfer phenomena. For processes in which electron or ionization transfer is involved, it is possible to indicate a mathematical framework of theory—but the employment of the terms and of the framework is still largely speculative.

Thus, the mathematical theory of 'primary' effects in the luminescence and chemical phenomena of radiation chemistry is only qualitatively established and only very moderately tested. It is only for processes which occur after 10^{-10} s that experiment may provide a test of the theory—but the theory involved is now that relating to the phenomena of diffusion kinetics. It can be tested only in the sense that numbers (e.g. for specific rates) derived on the basis of the kinetic scheme employed are reasonable and consistent with our entire body of knowledge—including, most particularly, those very special experiments contrived to go back almost to the origin of the effects.

The difficulty regarding excitation transfer processes in radiation chemistry is inherent in the time interval involved. The uncertainty principle itself may make it unmeaningful to attempt too intimate a search into the initial processes. Weaknesses in the theory of charge transfer may leave us with mathematical expressions, the terms of which are only qualitatively significant. Thus, as in any area of rapid scientific development, the radiation chemist is still constrained to speculation even though such speculation is becoming increasingly sophisticated.

REFERENCES

1. cf. B. K. Krotoszynski, *J. Chem. Phys.*, **41**, 2220 (1964).
2. cf. S. C. Lind, *Radiation Chemistry of Gases*, Reinhold, New York, 1961, p. 158.
3. cf. M. Burton, *Discussions Faraday Soc.*, **36**, 7 (1963).

4. H. A. Bethe and W. Heitler, *Proc. Roy. Soc. (London)*, **A146**, 83 (1934).
5. N. Bohr, *Phil. Mag.*, **25**, 10 (1913); *Det Danske Vied. Selsk. Mat. Fys. Medd.*, **18**, No. 8 (1948).
6. (a) F. Bloch, *Ann. Physik*, **16**, 285 (1933); cf. (b) M. S. Livingston and H. A. Bethe, *Rev. Mod. Phys.*, **9**, 263 (1937).
7. (a) H. A. Bethe, *Ann. Physik*, **5**, 325 (1930); *Ann. Physik*, **24**, 273 (1933); *Z. Physik*, **76**, 293 (1932); (b) U. Fano, *Ann. Rev. Nucl. Sci.*, **13**, 1 (1963); (c) J. L. Magee, *Ann. Rev. Phys. Chem.*, **12**, 389 (1961).
8. A. Mozumder and J. L. Magee, *Radiation Res.*, **28**, 203 (1966).
9. A. Dalgarno in *Atomic and Molecular Processes* (Ed. D. R. Bates), Academic Press, New York and London, 1962, Chapter 15.
10. F. Bloch, *Z. Physik*, **81**, 363 (1933).
11. J. F. Turner in *Studies in Penetration of Charged Particles in Matter*, Publication No. 1133, National Academy of Sciences, National Research Council, U.S.A., 1964.
12. E. Fermi, *Phys. Rev.*, **57**, 485 (1940).
13. H. Watanabe in *Penetration of Charged Particles in Matter*, Publication No. 752, National Academy of Sciences—National Research Council, U.S.A., 1960, pp. 152–162. Other references will be found in this summary article.
14. U. Fano in *Penetration of Charged Particles in Matter*, Publication No. 752, National Academy of Sciences—National Research Council, U.S.A., 1960, pp. 144–146.
15. G. Friedlander, J. W. Kennedy and J. M. Miller, *Nuclear and Radiochemistry*, (2nd Ed.), Wiley, New York, 1964, p. 89.
16. A. O. Allen, *The Radiation Chemistry of Water and Aqueous Solutions*, D. Van Nostrand, Princeton, 1961, Chapter 5. References to individual workers are given in this book.
17. A. Mozumder and J. L. Magee, *Radiation Res.*, **28**, 215 (1966).
18. L. C. Northcliffe, *Ann. Rev. Nucl. Sci.*, **13**, 67 (1963). See also L. C. Northcliffe in *Studies in Penetration of Charged Particles in Matter*, Publication No. 1133, National Academy of Sciences—National Research Council, U.S.A., 1964, pp. 173–186.
19. A. H. Samuel and J. L. Magee, *J. Chem. Phys.*, **21**, 1080 (1953).
20. A. K. Ganguly and J. L. Magee, *J. Chem. Phys.*, **25**, 129 (1956).
21. A brief summary is given here. For details of the method based on a Monte-Carlo approach, see A. Mozumder and J. L. Magee, *J. Chem. Phys.*, **45**, 3332 (1966).
22. L. Onsager, *Phys. Rev.*, **54**, 554 (1938).
23. M. Burton, *Molecular Crystals*, **4**, 61 (1968).
24. J. L. Magee, *Ann. Rev. Nucl. Sci.*, **3**, 171 (1953).
25. R. L. Platzman, *Radiation Res.*, **2**, 1 (1955).
26. G. J. Schulz, *Phys. Rev.*, **135**, A988 (1964); J. C. Y. Chen, *J. Chem. Phys.*, **40**, 3507, 3513; **41**, 3263 (1964); A. Herzenberg and F. Mandl, *Proc. Roy. Soc. (London)*, **A270**, 48 (1962).
27. J. Franck and E. Teller, *J. Chem. Phys.*, **6**, 861 (1938).
28. H. A. Schwartz, *Radiation Res. Suppl.*, **4**, 89 (1964).
29. R. L. Platzman, *Radiation Res.*, **17**, 419 (1962).

30. U. Fano in *Comparative Effects of Radiation* (Ed. M. Burton, J. S. Kirby-Smith and J. L. Magee), Wiley, New York, 1960, pp. 14–21.
31. J. Frenkel, *Phys. Rev.*, **37**, 17, 1276 (1931).
32. M. A. Dillon and M. Burton, in *Pulse Radiolysis* (Ed. M. Ebert, J. P. Keene, A. J. Swallow, J. H. Baxendale), Academic Press, New York, 1965, pp. 259–277.
33. K. Funabashi, P. J. Herley and M. Burton, *J. Chem. Phys.*, **43**, 3939 (1965).
34. J. B. Gallivan and W. H. Hamill, *J. Chem. Phys.*, **44**, 1279 (1966).
35. R. R. Hentz, D. B. Peterson, S. B. Srivastava, H. F. Barzynski and M. Burton, *J. Phys. Chem.*, **70**, 2362 (1966); R. R. Hentz, K. Shima and M. Burton, *J. Phys. Chem.*, **71**, 461 (1967).
36. J. L. Magee and K. Funabashi, *J. Chem. Phys.*, **34**, 1715 (1961).
37. Th. Förster, in *Comparative Effects of Radiation* (Ed. M. Burton, J. S. Kirby-Smith and J. L. Magee), Wiley, New York, 1960, p. 300.
38. R. Kubo and Y. Toyozawa, *Prog. Theor. Phys.*, **13**, 160 (1955).
39. J. Perrin and Mlle. Choucroun, *Compt. Rend.*, **184**, 1097 (1927); **189**, 1213 (1929).
40. H. Kallmann and M. Furst, *Phys. Rev.*, **79**, 857 (1950).
41. H. Dreeskamp and M. Burton, *Phys. Rev. Letters*, **2**, 45 (1959).
42. J. Yguerabide and M. Burton, *J. Chem. Phys.*, **37**, 1757 (1962).
43. J. D'Alessio and P. K. Ludwig, *IEEE Trans. Nucl. Sci.*, **12**, 351 (1965).
44. M. Burton, P. K. Ludwig, M. S. Kennard and R. J. Povinelli, *J. Chem. Phys.*, **41**, 2563 (1964).
45. H. Kallmann and F. London, *Z. Physik. Chem.*, **B2**, 207 (1928).
46. Th. Förster, *Ann. Physik*, **2**, 55 (1947).
47. D. L. Dexter, *J. Chem. Phys.*, **21**, 836 (1953).
48. R. Förster, *Bulletin No. 18, Division of Biology and Medicine*, U.S. Atomic Energy Commission (1965).
49. A. Y. Kurskii and P. S. Salivanenko, *Opt. Spectry.*, **8**, 340 (1960).
50. R. J. Povinelli, *Thesis*, University of Notre Dame, 1966.
51. M. Beer and H. C. Longuet-Higgins, *J. Chem. Phys.*, **23**, 1390 (1955).
52. M. Kasha, *Discussions Faraday Soc.*, **9**, 14 (1950).
53. M. V. Smoluchowski, *Z. Physik*, **17**, 557, 583 (1916).
54. Th. Förster, *Fluoreszenz Organischer Verbindungen*, Vandenhoeck and Ruprecht, Göttingen, 1951.
55. B. L. Sveshnikov, *Acta Physicochim. USSR*, **3**, 257 (1935).
56. J. Yguerabide, M. Dillon and M. Burton, *J. Chem. Phys.*, **40**, 3040 (1964).
57. R. Voltz, J. Klein, F. Heisel, H. Lauri, G. Laustriat and A. Coche, *J. Chim. Phys.*, **63**, 1259 (1966).
58. R. Voltz, G. Laustriat and A. Coche, *J. Chim. Phys.*, **63**, 1253 (1966).
59. S. Z. Toma and W. H. Hamill, *J. Am. Chem. Soc.*, **86**, 1478 (1964); J. P. Guarino and W. H. Hamill, *J. Am. Chem. Soc.*, **86**, 777 (1964).
60. P. K. Ludwig, *Molecular Crystals*, **4**, 147 (1968).
61. A. O. Allen and A. Hummel, *Discussions Faraday Soc.*, **36**, 95, 253 (1963); G. R. Freeman and J. M. Fajadh, *J. Chem. Phys.*, **43**, 86 (1965).
62. M. Burton and J. Yguerabide, *Proc. Conf. on Nuclear Electronics*, Belgrade Meeting of IAEA, May 1961, p. 71.
63. J. Franck and G. Cario, *Z. Physik*, **11**, 161 (1922).

64. W. West and B. Paul, *Trans. Faraday Soc.*, **28**, 688 (1932).
65. C. S. Schoepfle and C. H. Fellows, *Ind. Eng. Chem.*, **23**, 1396 (1931).
66. J. P. Manion and M. Burton, *J. Phys. Chem.*, **56**, 560 (1952).
67. M. Burton, *Discussions Faraday Soc.*, **36**, 7 (1963).
68. cf. K. Katsuura and M. Inokuti, *J. Chem. Phys.*, **41**, 989 (1964).
69. M. Burton, W. H. Hamill and J. L. Magee, *Proceedings of the Second Conference on the Peaceful Uses of Atomic Energy* (United Nations, Geneva, 1960), **29**, 391; J. Lamborn and A. J. Swallow, *J. Phys. Chem.*, **65**, 920 (1961); J. Bednar, *AEC Accession* No. 2375, Rept. No. UJV-1119/64 (1963); Z. Prasil, K. Vacek and J. Bednar, *Collection Czech. Chem. Commun.*, **30**, 2693 (1965); J. Bednar, *Collection Czech. Chem. Commun.*, **29**, 88 (1964); Z. Prasil, K. Vacek and J. Bednar, *AEC Accession* No. 35, 593, Rept. No. UJV-1075/64 (1964).
70. C. E. Klots, *J. Chem. Phys.*, **39**, 1571 (1963); C. E. Klots, *J. Chem. Phys.*, **44**, 2715 (1966).
71. M. Burton, J. Chang, S. Lipsky and M. P. Reddy, *Radiation Res.*, **8**, 203 (1958).
72. J. A. Stone and P. J. Dyne, *Radiation Res.*, **17**, 353 (1962); P. J. Dyne and W. M. Jenkinson, *Can. J. Chem.*, **38**, 539 (1960); P. J. Dyne and W. M. Jenkinson, *Can. J. Chem.*, **39**, 2163 (1961); P. J. Dyne and W. M. Jenkinson, *Can. J. Chem.*, **40**, 1746 (1962).
73. T. J. Hardwick, *J. Phys. Chem.*, **66**, 2132 (1962).
74. J. F. Merklin and S. Lipsky, *J. Phys. Chem.*, **68**, 3297 (1964).
75. E. A. Rojo and R. R. Hentz, *J. Phys. Chem.*, **69**, 3024 (1965).
76. G. Lunde and R. R. Hentz., *J. Phys. Chem.*, **71**, 863 (1967).
77. H. F. Barzynski, R. R. Hentz and M. Burton, *J. Phys. Chem.*, **69**, 2034 (1965).
78. J. R. Nash and W. H. Hamill, *J. Phys. Chem.*, **66**, 1097 (1962); see also W. Van Dusen, Jr. and W. H. Hamill, *J. Am. Chem. Soc.*, **84**, 3648 (1962); J. Roberts and W. H. Hamill, *J. Phys. Chem.*, **67**, 2446 (1963); J. A. Ward and W. H. Hamill, *J. Am. Chem. Soc.*, **87**, 1853 (1965).
79. G. Scholes and M. Simic, *Nature*, **202**, 895 (1964).
80. R. R. Hentz, D. B. Peterson, S. B. Srivastava, H. F. Barzynski and M. Burton, *J. Phys. Chem.*, **70**, 2362 (1966).
81. R. B. Cundall and P. A. Griffiths, *J. Am. Chem. Soc.*, **85**, 1211 (1963); *Discussions Faraday Soc.*, **36**, 111 (1963); *J. Phys. Chem.*, **69**, 1866 (1965); *Trans. Faraday Soc.*, **61**, 1968 (1965).
82. Ff. Williams, *J. Am. Chem. Soc.*, **86**, 3954 (1964); J. W. Buchanan and Ff. Williams, *J. Chem. Phys.*, **44**, 4377 (1966).
83. J. A. Stone and P. J. Dyne, *Can. J. Chem.*, **42**, 669 (1964).
84. E. A. Cherniak, E. Collinson and F. S. Dainton, *Trans. Faraday Soc.*, **60**, 1408 (1964).
85. D. B. Peterson, T. Arakawa, D. A. G. Walmsley and M. Burton, *J. Phys. Chem.*, **69**, 2880 (1965).
86. S. Arai and L. M. Dorfman, *J. Phys. Chem.*, **69**, 2239 (1965).
87. J. Nosworthy, *Trans. Faraday Soc.*, **61**, 1138 (1965).
88. J. P. Keene, E. J. Land and A. J. Swallow, *J. Am. Chem. Soc.*, **87**, 5284 (1965).
89. F. S. Dainton, T. J. Kemp, G. A. Salmon and J. P. Keene, *Nature*, **203**, 1050 (1964).

90. J. P. Keene, T. J. Kemp and G. A. Salmon, *Proc. Roy. Soc. (London)*, **A287**, 494 (1965).
91. R. R. Hentz, K. Shima and M. Burton, *J. Phys. Chem.*, **71**, 461 (1967).
92. R. R. Hentz, K. Shima and M. Burton, unpublished results.

Author Index

The numbers in square brackets refer to the reference numbers under which the Author's work is quoted in full at the end of the chapter.

Subject Index